Technical Aspects of Toxicological Immunohistochemistry

Syed A. Aziz • Rekha Mehta

Editors

Technical Aspects of Toxicological Immunohistochemistry

System Specific Biomarkers

 Springer

Editors
Syed A. Aziz
Regulatory Toxicology Research Division
Food Directorate
Health Products and Food Branch
Health Canada
Ottawa, ON, Canada

Department of Pathology
University of Ottawa
Ottawa, ON, Canada

Rekha Mehta
Regulatory Toxicology Research Division
Food Directorate
Health Products and Food Branch
Health Canada
Ottawa, ON, Canada

ISBN 978-1-4939-7965-3   ISBN 978-1-4939-1516-3   (eBook)
DOI 10.1007/978-1-4939-1516-3

Springer New York Heidelberg Dordrecht London

Printed on acid-free paper

Springer Science+Business Media LLC New York is part of Springer Science+Business Media (www.springer.com)

Preface

Immunohistochemistry and immunocytochemistry continue to play a critical role in cell and tissue biology and diagnostic pathology. The methodologies facilitate accurate analysis of the biochemistry of cells and tissues in relation to their morphological organization. In experimental toxicological pathology in particular, enhanced immunohistochemical techniques have evolved from a concept to a core tool over the past three decades. As additional procedures to the traditional histopathology, these techniques provide for the visualization of specific features in cells and tissues such as DNA adducts, proliferating cells, dying cells, stem cells, cell cycle markers, precancerous cells, hormone receptors, brain cell-specific receptors, and oncogenes, thus leading to a greater precision in the diagnosis of alterations in cells, tissues, and organs resulting from exposure to xenobiotics and dietary constituents.

The two books in this series have been made possible by contributions from leading experts in their respective fields. These books cover various immunohistochemical applications aimed at investigations utilizing in vivo rodent models including human xenograft tumor models, rodent and human cell lines, specimens of human lesions and tumors, and circulating tumor cells. They provide researchers in various subdisciplines of toxicology and pathologists with a review of available methods and approaches for staining in various target tissues, including an understanding of the technical and tissue-, antigen-, and antibody-specific issues that may be encountered during the staining procedure.

This book, *Technical Aspects and Applications of Toxicological Immunohistochemistry: System Specific Biomarkers*, comprises 11 chapters with color micrographs and descriptions relating to normal tissue biology and altered pathophysiology in response to toxic environments and/or disease. The subjects covered include tissue microarray technology, fluorescent in situ hybridization, image analysis methodology, and immunochemical applications for oncology drug development and for assessment of responses to toxicants in the male and female reproductive systems, the circulatory system, lymphoid tissue, bone, and skeletal muscle. The information in this book is rounded out by 11 chapters in the companion book, *Toxicological Immunohistochemistry: Mechanistic and Predictive*

Biomarkers, which examines immunochemical markers that may be used for target organ-specific screening of toxicity-related responses to evaluate or predict a chemical's mode of action.

The described works benefit scientists and health professionals in industry, government, and academia and students in pathology as well as in various health disciplines including food, nutritional, drug, clinical, and environmental toxicology.

The editors gratefully thank all the participating authors without whose contributions the scope of this book would have been very limited. We also acknowledge the many scientists whose cited references provide evidence of their significant contributions toward the advancement of the field of enhanced immunohistochemistry. We further thank our scientific colleagues in the Bureau of Chemical Safety, Food Directorate, Health Canada, for their suggestions and in-kind contributions in the course of preparing this publication. Finally, we dedicate this work to our respective families in recognition of their endurance and continual support during our research endeavors.

Ottawa, ON, Canada   Syed A. Aziz
                                                                                        Rekha Mehta

Contents

Contributors

Filomena Adega University of Trás-os-Montes and Alto Douro, Department of Genetics and Biotechnology (DGB), Laboratory of Cytogenomics and Animal Genomics (CAG), Vila Real, Portugal

Syed A. Aziz, PhD Regulatory Toxicology Research Division, Bureau of Chemical Safety, Food Directorate, Health Products and Food Branch, Health Canada, Ottawa, ON, Canada

Department of Pathology and Laboratory Medicine, Faculty of Medicine, University of Ottawa, Ottawa, ON, Canada

Paulo Sergio Bossini, PT, MSc, PhD Physical Therapy Department, Federal University of São Carlos, São Carlos, SP, Brazil

Raquel Chaves University of Trás-os-Montes and Alto Douro, Department of Genetics and Biotechnology (DGB), Laboratory of Cytogenomics and Animal Genomics (CAG), 5001-801, Vila Real, Portugal

University of Lisboa, Faculty of Sciences, BioISI– Biosystems & Integrative Sciences Institute, Campo Grande, Lisboa, Portugal

Daniel G. Cyr INRS-Institut Armand-Frappier, University of Quebec, Laval, QC, Canada

Patrick J. Devine Novartis Institutes for BioMedical Research, Inc., Cambridge, MA, USA

E. Driskell, DVM, PhD, Diplomate ACVP Department of Pathobiology, University of Illinois, Urbana, IL, USA

Veterinary Diagnostic Laboratory, University of Illinois, Urbana, IL, USA

Michael R. Hamblin, PhD Wellman Center for Photomedicine, Massachusetts General Hospital, Boston, MA, USA

Department of Dermatology, Harvard Medical School, Boston, MA, USA

Harvard-MIT Division of Health Sciences and Technology, Cambridge, MA, USA

Kamsiah Jaarin Faculty of Medicine, Department of Pharmacology, The National University of Malaysia, Kuala Lumpur, Malaysia

Fang Jiang Cancer Discovery, Global Pharmaceutical Research and Development, Abbott Laboratories, Abbott Park, IL, USA

Yusof Kamisah Faculty of Medicine, Department of Pharmacology, The National University of Malaysia, Kuala Lumpur, Malaysia

Gian Kayser Institute of Pathology, University of Freiburg, Freiburg, Germany

Klaus Kayser Institute of Pathology, Charite, Berlin, Germany

Evelyn M. McKeegan Cancer Discovery, Global Pharmaceutical Research and Development, Abbott Laboratories, Abbott Park, IL, USA

Rekha Mehta Regulatory Toxicology Research Division, Bureau of Chemical Safety, Food Directorate, Health Products and Food Branch, Health Canada, Ottawa, ON, Canada

Ana Mendes-da-Silva University of Trás-os-Montes and Alto Douro, Department of Genetics and Biotechnology (DGB), Laboratory of Cytogenomics and Animal Genomics (CAG), Vila Real, Portugal

University of Lisboa, Faculty of Sciences, BioISI– Biosystems & Integrative Sciences Institute, Campo Grande, Lisboa, Portugal

Ali Mobasheri, BSc, ARCS(Hons), MSc The D-BOARD European Consortium for Biomarker Discovery, School of Veterinary Medicine, Faculty of Health and Medical Sciences, University of Surrey, Guildford, Surrey, UK

Center of Excellence in Genomic Medicine Research (CEGMR), King Abdul Aziz University, Jeddah, Kingdom of Saudi Arabia

Isa Naina Mohamed Bone Metabolism Research Group, Department of Pharmacology, Faculty of Medicine UKM, Kuala Lumpur, Malaysia

Norazlina Mohamed Bone Metabolism Research Group, Department of Pharmacology, Faculty of Medicine UKM, Kuala Lumpur, Malaysia

Norliza Muhammad Bone Metabolism Research Group, Department of Pharmacology, Faculty of Medicine UKM, Kuala Lumpur, Malaysia

Chun-Yi Ng Faculty of Medicine, Department of Pharmacology, The National University of Malaysia, Kuala Lumpur, Malaysia

Nivaldo A. Parizotto, PT, MSc, PhD Physical Therapy Department, Federal University of São Carlos, São Carlos, SP, Brazil

Harvard Medical School, Boston, MA, USA

Isabelle Plante INRS-Institut Armand-Frappier, University of Quebec, Laval, QC, Canada

Elvy Suhana Mohd Ramli Bone Metabolism Research Group, Department of Pharmacology, Faculty of Medicine UKM, Kuala Lumpur, Malaysia

Natalia C. Rodrigues, PT, MSc, PhD Physical Therapy Department, Federal University of São Carlos, São Carlos, SP, Brazil

Ahmad Nazrun Shuid Bone Metabolism Research Group, Department of Pharmacology, Faculty of Medicine UKM, Kuala Lumpur, Malaysia

Kuldeep Singh, DVM, MS, PhD, Diplomate ACVP Department of Pathobiology, University of Illinois, Urbana-Champaign, Urbana, IL, USA

Veterinary Diagnostic Laboratory, University of Illinois, Urbana-Champaign, Urbana, IL, USA

A. Stern, DVM, CMI-IV, CFC, Diplomate ACVP Veterinary Diagnostic Laboratory, University of Illinois, Urbana-Champaign, Urbana, IL, USA

Chapter 1
Immunohistochemistry: Overview, Its Potential, and Challenges

Syed A. Aziz and Rekha Mehta

Overview

Immunohistochemistry (IHC) blends anatomical, immunological, as well as bio-chemical disciplines to detect discrete tissue constituents by the capture of target antigens with characteristic antibodies tagged with an appropriate label. IHC makes it possible to visualize the distribution and localization of characteristic cellular elements within cells and in the proper tissue context. The concept of IHC has been known since the 1930s, but it was not until 1942 that the first IHC study was reported. Coons et al. [1] used FITC-labeled antibodies to determine Pneumococcal antigens in infected tissue. Since then, corrections and advances have been made in protein conjugation, tissue fixation methods, detection labels, microscopy, and image analysis, making IHC a routine and essential tool in diagnostic and research laboratories.

S.A. Aziz, Ph.D. (✉)
Regulatory Toxicology Research Division, Bureau of Chemical Safety, Food Directorate,
Health Products and Food Branch, Health Canada, Sir Frederick Banting Research Institute,
D-401 Banting, Building, Postal Locator 2202D, 251 Sir Frederick Banting Driveway,
Tunney's Pasture, Ottawa, ON, Canada, K1A 0K9

Department of Pathology and Laboratory Medicine, University of Ottawa,
Ottawa, ON, Canada
e-mail: syed.a.aziz@hc-sc.gc.ca

R. Mehta
Regulatory Toxicology Research Division, Bureau of Chemical Safety, Food Directorate,
Health Products and Food Branch, Health Canada, Sir Frederick Banting Research Institute,
D-401 Banting, Building, Postal Locator 2202D, 251 Sir Frederick Banting Driveway,
Tunney's Pasture, Ottawa, ON, Canada, K1A 0K9

© Springer Science+Business Media New York 2016
S.A. Aziz, R. Mehta (eds.), *Technical Aspects of Toxicological Immunohistochemistry*, DOI 10.1007/978-1-4939-1516-3_1

The Potential of IHC

Presently IHC is used clinically for disease diagnosis and prognosis, therapeutic research and drug development, and biological/toxicological research. For example, using specific tumor markers, physicians use IHC to diagnose a cancer as benign or malignant, determine the stage and grade of a tumor, and identify the cell type and origin of a metastasis to find the site of the primary tumor, thus providing the diagnosis along with prognosis [2–4]. In experimental toxicological pathology in particular, enhanced IHC techniques have evolved from a theory to a core tool over the past three decades. As described in various chapters of this book, IHC is now a routine additional procedure to the traditional histopathology in many laboratories, leading to the visualization of specific features such as proliferating cells, dying cells, stem cells, cell surface antigens, cell cycle and differentiation related markers, gap-junction related proteins, precancerous cells, and hormone and brain cell specific receptors, thus allowing a greater precision in the diagnosis of target cell-specific alterations resulting from chemical exposure.

From single usually brown staining of IHC to visualize the reaction, this technique has further evolved into a multiple staining technique for looking at two or more antigens in the same tissue section simultaneously. While on one hand such advances make the technical aspects more challenging, the benefits result in shorter times to diagnosis, improved inter- and intraobserver reproducibility in the diagnosis of benign lesions, and an improved understanding regarding disease stage-specific concurrent expression of multiple antigens, leading to better informed treatment options [5]. For example, simultaneous multiple staining with ADH-5 cocktail containing five breast cancer implicated antigens (cytokeratins 5, 14, 7, 18, and p63) is now routinely used in the clinic with improved diagnostic consistency to aid the pathological differentiation of usual ductal hyperplasia from atypical ductal hyperplasia or low-grade ductal carcinoma in situ and for ruling out microinvasion and distinguishing invasive carcinoma from pseudoinvasive lesions and for follow-up patient treatment decisions [5, 6].

Challenges

For the basic theory, reagents, and protocols of the IHC staining procedure, the reader is referred to an excellent manual published by Dako [7] and the references therein. While some of the chapters in our current publication provide details of the applied protocols, these are aimed at particular toxicologic pathology applications and hence refer to specific tissues and antigens of interest. Therefore, any IHC application for a new antigen may require modification of the described protocols.

One of the major challenges in adopting IHC in any lab has been the consideration of quality control (QC) to ensure reproducibility of the staining procedure for accurate identification and interpretation of levels of antigens for biological and molecular characterization of tissue specimens. This is particularly important in a

Table 1.1 The "Total Test" approach for standardization of IHC tests in a clinical setting (adapted from Taylor [8, 11])

IHC stage	Step
Preanalytic	Test selection Specimen type Tissue acquisition, pre-fixation/transport time Fixation, type and time Processing, temperature
Analytic	Antigen retrieval procedure Selection of primary Ab's Protocol, labeling agents Reagent variation Control selection Technician training/certification Lab certification/QA programs
Postanalytic	Assessment of control performance Description of results Interpretation/reporting Pathologist, experience and CME specific to IHC

clinical lab where IHC data are used to provide additional diagnostic, predictive, and prognostic information to the pathologist and the clinician for appropriate patient treatment. Thus, a concern regarding a lack of standardization of the IHC procedure in the clinical setting for Her 2 IHC assay led to a series of recommendations by experts from the Food and Drug Administration and the National Institute of Standards and Technology. A "Total Test" approach to consider all aspects of the IHC procedure, including a need for universal control materials (or reference standards), was suggested when standardizing IHC assays [8–10]. The "Total Test" approach thus involves standardization along all three stages of the IHC procedure and for each of several steps within each stage (Table 1.1).

Unlike the clinical setting, there is no published QC and assurance guideline specific to a toxicological pathology laboratory according to our literature search. As will become obvious in ensuing chapters, this is most likely because as in the case of any nonroutine test in a research setting, each antigen detection method requires its own in-built QC and assurance as a result of variations in available antibodies and the cell/ tissue types being studied for antigen localization. However, once the IHC method is established for a specific antigen, it is important that some type of QC procedure is in place to ensure consistent data. Therefore, based on the recommendations for QC and assurance in clinical IHC, we suggest that the "Total Test" approach be also considered as some form of QC for an IHC procedure in a toxicological pathology lab within a research setting since the same steps shown in Table 1.1 would be required for the IHC test. Along the same lines, the guideline for human tissues provided by the Canadian Association of Pathologists [12] is another one that could be adapted to suit toxicology needs and for rodents or other mammalian tissues since almost all the antibodies so far available for use in humans are also available and compatible for use in rodents.

Conclusion

IHC and immunocytochemistry continue to play a critical role in cell and tissue biology as diagnostic aids in pathology and toxicology despite the QC challenges we have mentioned. As the following chapters will demonstrate, the advancing methodologies facilitate accurate analysis of the biochemistry of many diverse cells and tissues in relation to their morphological organization.

References

1. Coons AA, Creech HJ, Norman RJ, Ernst B. The demonstration of pneumococcal antigen in tissues by the use of fluorescent antibody. J Immunol. 1942;45:159–70.
2. Elias JM. Immunohistochemical methods. In: Immunohistopathology: a practical approach to diagnosis. 2nd ed. Chicago, IL: ASCP; 2003. p. 1–110.
3. Taylor CR, Shi S-R, Barr NJ, et al. Techniques of immunohistochemistry: principles, pitfalls, and standardization. In: Dabbs D, editor. Diagnostic immunohistochemistry. 2nd ed. New York: Elsevier; 2006. p. 1–42.
4. Taylor CR, Cote RJ. Immunomicroscopy: a diagnostic tool for the surgical pathologists, Major problems in pathology, vol. 19. 3rd ed. Philadelphia, PA: Saunders; 2005. p. 26–8.
5. Jain RK, Mehta R, Dimitrov R, Larsson LG, Musto PM, Hodges KB, Ulbright TM, Hattab EM, Agaram N, Idrees MT, Badve S. Atypical ductal hyperplasia: interobserver and intraobserver variability. Mod Pathol. 2011;24:917–23.
6. Yeh IT, Mies C. Application of immunohistochemistry to breast lesions. Arch Pathol Lab Med. 2008;132:349–58.
7. National Committee for Clinical Laboratory Standards. Quality assurance for immunocytochemistry; approved guideline. NCCLS document MM4-AC. Wayne, PA: National Committee for Clinical Laboratory Standards; 1999.
8. Taylor CR. The total test approach to standardization of immunohistochemistry. Arch Pathol Lab Med. 2000;124:945–51.
9. Goldstein NS, Hewitt SM, Taylor CR, Members of Ad-Hoc Committee on Immunohistochemistry Standardization, et al. Recommendations for improved standardization of immunohistochemistry. Appl Immunohistochem Mol Morphol. 2007;15:124–33.
10. Kumar GL, Rudbeck L. Education guide—immunohistochemical staining methods. 5th ed. Carpenteria, CA: Dako North America; 2009. p. 1–154.
11. Taylor CR. Immunohistochemical standardization and ready-to-use antibodies. In: Kumar GL, Rudbeck L, editors. Education guide—immunohistochemical staining methods. 5th ed. Carpenteria, CA: Dako North America; 2009. p. 21–8.
12. Torlakovic EE, Riddell R, Banerjee D, El-Zimaity H, et al. Canadian Association of Pathologists–Association canadienne des pathologistes National Standards Committee/ Immunohistochemistry: best practice recommendations for standardization of immunohistochemistry tests. Am J Clin Pathol. 2010;133:354–65.

Chapter 2
Tissue Microarray Technology and Its Potential Applications in Toxicology and Toxicological Immunohistochemistry

Ali Mobasheri

Introduction

The completion and annotation of the Human Genome Project (HGP)[1] in 2003 has provided the basic structural information on all human genes. The project was coordinated by the U.S. Department of Energy[2] and the National Institutes of Health (NIH).[3] During the early years of the HGP, the Wellcome Trust (U.K.)[4] became a major partner and later additional contributions joined the project from Japan, France, Germany, China, and others. The HGP took nearly 13 years to complete. Its completion coincided with the 50th anniversary of Watson and Crick's description of the primary structure of DNA. During the HGP's 13 years significant advances in computer technology and bioinformatics resulted in accelerated progress. In June 2000, the rough draft of the human genome was completed a year ahead of schedule. In February 2001, the working draft was completed. The primary aims of the HGP were to identify all the approximately 20,000–25,000 genes in the human genome, determine the sequences of all the genes in the human genome, and store this information in publicly accessible databases. Although the HGP has been

---

[1] http://www.ornl.gov/sci/techresources/Human_Genome/home.shtml

[2] http://energy.gov/

[3] http://www.nih.gov/

[4] http://www.wellcome.ac.uk/

A. Mobasheri, BSc ARCS (Hons), MSc (✉)
The D-BOARD European Consortium for Biomarker Discovery, School of Veterinary Medicine, Faculty of Health and Medical Sciences, University of Surrey, Duke of Kent Building, Guildford, Surrey GU2 7XH, UK

Center of Excellence in Genomic Medicine Research (CEGMR), King Abdul Aziz University, Jeddah 21589, Kingdom of Saudi Arabia
e-mail: a.mobasheri@surrey.ac.uk; http://www.surrey.ac.uk/vet/people/ali_mobasheri/index.htm

© Springer Science+Business Media New York 2016
S.A. Aziz, R. Mehta (eds.), *Technical Aspects of Toxicological Immunohistochemistry*, DOI 10.1007/978-1-4939-1516-3_2

completed, the analysis of the data is expected to continue for many years. This includes functional annotation of the genes, improving bioinformatic tools for data analysis, transferring related technologies to the private sector, and addressing the ethical, legal, and social issues (ELSI) that arise from the project. The HGP has also contributed to the development of the genome projects of several other organisms. The HGP combined with the ongoing efforts to sequence and annotate the genomes of common experimental animals, companion animals, domesticated farm animals, fish, and other veterinary species (National Animal Genome Research Program[5]) will provide us with a much better understanding of the function of individual genes and the role of gene families in normal physiology, pathophysiology, and pathology.

It is important to bear in mind that the HGP and animal genome projects are essentially the tip of a very large and deep iceberg and only the beginning of a larger quest aimed at understanding the physiology of humans and animals. Genes merely provide the biological "script" for life; it is the protein products of genes that are the real "actors," performing specific roles in cells and accomplishing the functional objectives of the genome. The 20,000–25,000 genes in the human genome are likely to give rise to hundreds of thousands of different proteins; further levels of complexity are provided by the proteome (the protein complement of a cell) and the multitude of posttranslation modifications that are possible [1]. Numerous types of posttranslational modification (i.e., glycosylation, acylation, limited proteolysis, phosphorylation, isoprenylation) that individual proteins may undergo to modulate their function will add further levels of sophistication and complexity. The recognition that higher levels of complexity exist over and above the genome has resulted in new terminologies such as the "physiomics" and "metabolomics," all of which lead to "pathomics" and new multidisciplinary projects like the "physiome project" [2, 3].

The Emergence and Dominance of Post-genomic Technologies

The development and application of high-throughput post-genomic screening techniques has fundamentally transformed biomedical research. Large-scale academic and industrial efforts are presently under way to apply genomics and various related post-genomic applications, including proteomics for the identification of targets for new diagnostics and therapeutics (Fig. 2.1). Identification, selection, validation, and prioritization of the best targets from tens of thousands of candidate genes and corresponding proteins are daunting tasks that require application of multiple high-throughput techniques. Proteomics and associated high-throughput technologies are being developed to offer the scientific community a better understanding of normal biology and physiology of cells, tissues, and organs and to gain a better appreciation of the pathophysiology of common diseases in humans and animals [4]. Detailed proteomic and immunohistochemical analysis of molecular targets in situ at the cellular and tissue level, quantitative assessment of their relative expression levels

---

[5] http://www.animalgenome.org/

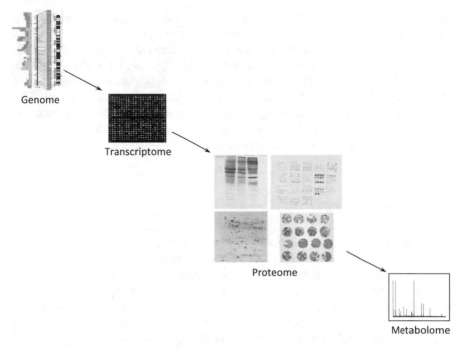

Fig. 2.1 Hierarchical display of genomic, transcriptomic, proteomic, and metabolomic technologies in biomedical research. TMA technology is an important component of proteomics for high-throughput studies of protein biomarkers

across all tissues and diseases, and evaluation of their clinical significance would therefore provide significant additional information and help prioritize target selection. Identification of novel biomarkers for early disease detection, prediction, and prognosis will reduce the morbidity and mortality associated with diseases and help define new therapeutic targets, drugs, and vaccines [4].

The Proliferation of Microarray Technologies

The development of high-throughput screening technologies has significantly proliferated in the last decade [5–9]. It is now possible to analyze the expression of thousands of genes at once from small quantities of tissue or cell samples [10]. Microarray technologies have pervaded all the medical, veterinary, and biological disciplines. These technologies allow investigators to determine the effects of hormones, cytokines, drugs, and environmental conditions on gene expression levels. The up- or downregulation of subsets of genes in response to the stimulation of interest is then compared with un-stimulated controls. Microarray technology can also be used to help understand cancer development, to improve patient treatment and management, and to identify those predisposed to develop cancer [11, 12]. Microarray expression patterns are currently being integrated with clinical data to identify new biomarkers to predict the potential aggressive behavior of tumors. Protein arrays are also becoming increasingly popular as the equivalent tool in the hands of the proteomics expert. However, proteomic approaches have received less interest from histopathologists and pathologists who are primarily interested in equivalent microscopic and histologic technologies, which will allow them to screen large numbers of tissue specimens for protein expression information and for discovering diagnostic and prognostic correlations.

Tissue Microarrays

Tissue microarray (TMA) technology is a relatively new technique, which allows rapid and simultaneous visualization of molecular targets in hundreds or thousands of tissue specimens [13] (see Fig. 2.2). The molecular targets can be DNA, RNA, or protein depending on the type of probe used for the analysis [14]. This methodology was established in response to an acute need for a reliable high-throughput technique for comprehensive screening of large numbers of human tumors [12, 13, 15]. Figure 2.3 illustrates the steps involved in producing TMAs. High-throughput TMA

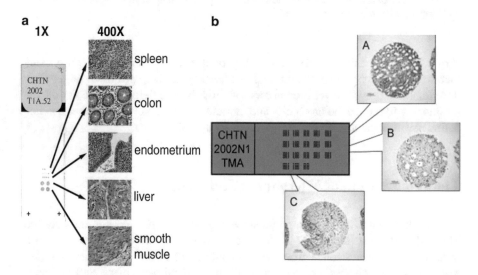

Fig. 2.2 (**a**) Test microarrays of normal human tissues provide researchers with an inexpensive TMA of a limited number of formalin-fixed paraffin-embedded samples that can be used for optimization of immunohistochemical and in situ hybridization parameters prior to use of more comprehensive tissue microarrays. (**b**) Example of a high-density CHTN showing hematoxylin and eosin stained cores from normal human kidney cortex (**A**), medulla (**B**), and prostate (**C**)

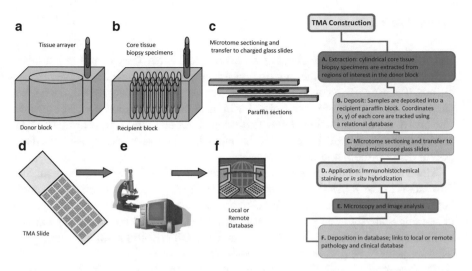

Fig. 2.3 TMA design and construction. The use of high-density TMAs allows validation of novel protein biomarker discovery and validation across all human tissues by associating TMA data with clinical and pathological data. Using this approach, novel markers of disease pathogenesis may be validated

technology enables the clinical relevance of molecular markers to be assessed simultaneously in multiple tissue specimens [14]. The widespread adoption of TMAs in research laboratories worldwide has replaced the conventional and frustratingly tedious one-slide-one-section approach, in which individual archival clinical specimens were placed on separate microscope slides, with the ability to assess RNA, DNA, or protein biomarker expression in numerous individual tissue specimens in a single carefully controlled experiment [14]. This technology has found applications in almost all areas of human oncology and clinical medicine, and is significantly accelerating the pace of advances in translational research through more efficient assessment of new biomarkers. This technique has not only allowed pathologists to reevaluate the use of existing tumor markers but has also facilitated clinical practice by linking outcome and response to chemotherapy.

Basic scientists and clinical investigators have been using multiple tissue Northern and Western blots for many years for studies of gene and protein expression. However, these methods are expensive, laborious, and time-consuming to establish. More importantly these conventional approaches do not permit high-throughput analysis. Proteomics and TMAs are receiving considerable interest from scientists who are increasingly embracing these new technologies. The aim of this chapter is to introduce the readers to the concept of TMAs and discuss the existing and potential applications of TMA technology in toxicological immunohistochemistry. Despite its impact on other areas of research, TMA has not yet gained recognition in the area of toxicology. TMA technology is likely to have an increasingly important role in toxicology research and it is hoped that this chapter is to bring toxicologists up to date with the basic principles and future capabilities of TMAs.

What Are TMAs?

TMAs are an ordered array of tissue cores on a glass slide. They permit immunohistochemical analysis of numerous tissue sections under identical experimental conditions. The arrays can contain samples of every organ in the human body, or a wide variety of common tumors and obscure clinical cases alongside normal controls. The arrays can also contain pellets of cultured tumor cell lines. These arrays may be used like any histological section for immunohistochemistry and in situ hybridization to detect protein and gene expression.

The TMA technique was originally described in 1987 by Wan, Fortuna, and Furmanski in the *Journal of Immunological Methods* [16]. They published a modification of Battifora's "sausage" block technique whereby tissue cores were placed in specific spatially fixed positions in a block [17]. The technique was popularized by Kononen and colleagues in the laboratory of Ollie Kallioneimi after they published a paper in Nature Medicine in 1998 [15].

This technology will allow investigators to analyze numerous biomarkers over essentially identical samples, develop novel prognostic markers, and validate potential drug targets. The ability to combine TMA technology with DNA microarrays and proteomics makes it a very attractive tool for the analysis of gene expression in clinically stratified tumor specimens and to relate the expression of each particular protein with clinical outcome. Public domain software allows researchers to examine digital images of individual histological specimens from TMAs, evaluate and score them, and store the quantitative data in a relational database. TMA technology may be specifically applied to the profiling of proteins and biomarkers of interest in other pathophysiological conditions such as congestive heart failure, renal disease, hypertension, diabetes, cystic fibrosis, and neurodegenerative disorders. This technology has the capacity to allow quantitative assessment of changes in protein markers across a range of tissues samples from multicenter toxicological studies. However, thus far, very few toxicologists have taken advantage of this technology.

Why Use TMAs?

A classic problem in biology is validation of a new concept on actual human tissue. A frequent problem is there is not enough tissue, either in amount or case number, to complete the analysis. This chronic shortage of tissue is partly due to the fact that the initial sampling is typically done by a pathologist whose primary goal is to provide a diagnostic consultation, not material for future experimentation. The difficulty of working with human tissue is further complicated by increasingly stringent guidelines regarding obtaining informed consent for its use. Thus, when tissue is obtained, it should be optimally managed to maximize its value. Tissue microarray represents a mechanism for highly effective use of this scarce resource.

In terms of benefits TMA technology offers numerous advantages including amplification of a scarce resource (i.e., rare archival tissue specimens; monoclonal or polyclonal antibodies that may no longer be available), experimental uniformity, decreased assay volume and use of antibodies, and significantly reduced or eliminated slide-to-slide variation.

How Is TMAs Produced?

TMAs are produced by re-locating tissue from conventional histologic paraffin blocks such that tissue from multiple patients or blocks can be seen on the same slide when the TMA paraffin block is sectioned. This is done by using a needle to biopsy a standard histologic section and placing the core into an array on a recipient paraffin block.

A standard histologic section is about 3–5 mm thick, with variation depending on the submitting pathologist or tech. After use for primary diagnosis, the sections can be cut 50–100 times depending on the care and skill of the sectioning technician. Thus, on average, each archived block might yield material for a maximum of 100 assays. If this same block is processed for optimal microarray construction, it could routinely be needle biopsied 200–300 times or more depending on the size of the tumor in the original block (theoretically it could be biopsied 1000s of times based on calculations of area, but empirically, 200–300 is selected as a conservative estimation) Then, once tissue microarrays are constructed, they can be judiciously sectioned in order to maximize the number of sections cut from an array. The sectioning process uses a tape-based sectioning aid (from Instrumedics Inc.[6]) that allows cutting of thinner sections. Optimal sectioning of arrays is obtained with about 2–3 μm sections. Thus, instead of 50–100 conventional sections or samples for analysis from one tissue biopsy, the microarray technique could produce material for 500,000 assays (assuming 250 biopsies per section times 2000 2.5 μm sections per 5 mm array block) represented as 0.6 mm disks of tissue. Thus, this technique essentially amplifies (up to 10,000-fold) the limited tissue resource.

Using this technology, each tissue sample is treated in an identical manner. Like conventional formalin-fixed paraffin-embedded material, tissue microarrays are amenable to a wide range of techniques including histochemical stains, immunologic stains with either chromogenic or fluorescent visualization, in situ hybridization (including both mRNA ISH and FISH), and even tissue microdissection techniques. For each of these protocols, conventional sections can have substantial slide-to-slide variability associated with processing 300 slides (e.g., 20 batches of 15 slides). The tissue microarrays allow the entire cohort to be analyzed in one batch on a single slide. Thus, reagent concentrations are identical for each case, as are incubation times and temperatures, wash conditions, and antigen retrieval, if necessary. Another significant advantage is that only a very small (a few μl) amount of antibody reagent is

---

[6] http://www.instrumedics.com/

required to analyze an entire cohort. This advantage raises the possibility of use of tissue microarrays in screening procedures (e.g., in hybridoma screening), a protocol that is impossible using conventional sections. It also saves money when reagents are costly. Finally, there are occasions where the original block must be returned to the donating institution. In these cases, the block may be cored a few times without destroying the block. Then upon subsequent sectioning, it is still possible to make a diagnosis, even though tissue has been taken for array-based studies.

TMA Resources

In this section, TMAs and TMA services of publicly funded noncommercial and commercial sources are briefly summarized and discussed. Although the author has used commercial TMAs in one of his research papers [18], he does not have any affiliations with commercial TMA manufacturers. In most of the publications arising from our research laboratory, we have used noncommercial TMAs. We have also used custom-made low-density TMAs in studies of aquaporin water channel expression across kidneys of several mammalian species [18–22].

CHTN TMAs

The Cancer Diagnosis Program of the National Cancer Institute (NCI[7]) initiated the Cooperative Human Tissue Network (CHTN[8,9]) in 1987 to provide increased access to human tissue for basic and applied scientists from academia and industry to accelerate the advancement of discoveries in cancer diagnosis and treatment. The CHTN is a unique resource that provides remnant human tissues and fluids from routine procedures to investigators who utilize human biospecimens in their research. The CHTN comprises five adult divisions and one pediatric division. Each of the adult divisions coordinates investigator applications/requests based upon the investigator's geographic location within North America. The Pediatric Division manages all investigators who request pediatric specimens only. The CHTN divisions share coordination for requests from outside North America.

Unlike tissue banks, the CHTN works prospectively with each investigator to tailor specimen acquisition and processing to meet their specific project requirements. Because the CHTN is funded by the NCI, the CHTN is able to maintain nominal processing fees for its services. The purpose of CHTN is to provide biomedical researchers access to human tissue specimens by offering tissue microarrays (TMAs) of neoplastic and non-neoplastic human tissues (see Fig. 2.4). The CHTN

---

[7] http://www.cancer.gov/

[8] http://chtn.nci.nih.gov/

[9] http://faculty.virginia.edu/chtn-tma/

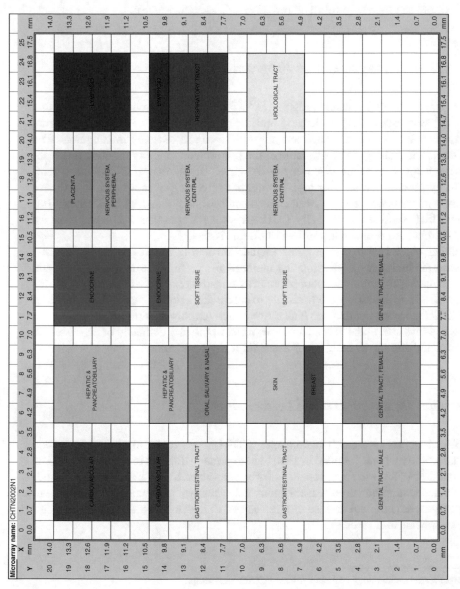

Fig. 2.4 Tissue microarray of formalin-fixed paraffin-embedded samples that includes 66 normal tissue types present in the human body. This microarray which is available from the Cooperative Human Tissue Network (CHTN) is codenamed CHTN2002N1. For a list of tissues corresponding to the codes, see the CHTN website (http://www.chtn.nci.nih.gov/)

divisions work both independently with individual investigators and together as a seamless unit to fulfill requests that are difficult to serve by any single division. The CHTN's unique informatics system allows each division to effectively communicate and network the needs of its investigators to all CHTN divisions. The Network as a whole can then help fulfill an investigator's specific request.

The Tissue Array Research Program

The Tissue Array Research Program (TARP) [10] at the National Cancer Institute (NCI)[11] was established to develop and distribute multi-tumor tissue microarray slides (see Fig. 2.5) and related technologies to cancer research investigators. The rationale for making TMA available to the research community was that this technology helps to expedite discovery of the novel targets important in cancer treatment by providing a tool for high-throughput screening of multiple tumor tissues using immunohistochemical, in situ hybridization, and fluorescent in situ hybridization (FISH) analyses. Essentially TARP arrays were established as novel tools for high-throughput molecular profiling of tumor tissues.

Other objectives of the program include furthering the development of tissue microarray technology, investigating methods for probing tissue microarrays, and disseminating information about microarray construction and use. A variety of different tissue microarrays are available to researchers depending on what material the TARP Lab can obtain from NCI tissue banks for construction of the arrays. The TARP arrays are updated frequently to reflect both the demands of the research community and the availability of tissue for array construction.

National Disease Research Interchange

National Disease Research Interchange (NDRI)[12] is a nonprofit, federally funded organization that was founded in 1980 to help procure difficult to obtain tissues. The aim of the NDRI is to serve scientists with customized biomaterials for use in studies to understand human disease. Human cells, tissues, and organs are available to investigate how human disease progresses and to develop new drugs and therapies for treatments and cures.

---

[10] https://ccrod.cancer.gov/confluence/display/CCRTARP/Home

[11] http://www.cancer.gov/

[12] http://www.ndriresource.org/

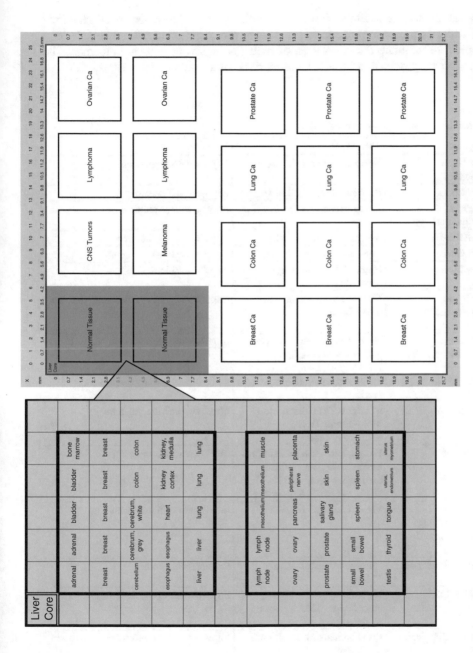

Fig. 2.5 Illustration of a TARP multi-tumor tissue microarray originally available from the National Cancer Institute which contains normal and tumor specimens on a single slide. Details and orientation of normal tissues are also represented on the enlarged panel on the left of the main TARP 4 microarray slide. For further details, see the TARP website (http://www.tarp.nih.gov/)

Commercially Produced TMAs

TMAs are widely available from commercial sources. However, most pathology laboratories prefer to make in-house arrays from their often extensive pathology archive to facilitate the correlation of their findings with clinical outcomes. The information provided in Table 2.1 summarizes some commercial sources of TMAs.

Table 2.1 Commercial sources of TMAs

Company	TMA types	Website
OriGene Technologies	Normal and cancer TMAs	http://www.origene.com/
BioChain	Paraffin and frozen TMAs	http://www.biochain.com/
Insight Biotechnology	Breast carcinoma TMAs	http://www.insightbio.com/
IMGENEX	Normal and cancer TMAs	http://www.imgenex.com/
BioGenex	Low- and high-density TMAs	http://www.biogenex.com/
US Biomax	Human TMAs and single tissue slides	http://www.biomax.us
Protein Biotechnologies Inc.	Human TMAs and human tumor lysates	http://www.proteinbiotechnologies.com/
LifeSpan BioSciences, Inc.	Normal and tumor TMAs as well as cardiovascular, neurological, inflammatory, metabolic, and toxicology products	http://www.lsbio.com/
Invitrogen	MaxArrayTM Human Multi-Tumor Tissue Microarray Slides	http://www.invitrogen.com/
Asterand	Frozen, fixed, and fresh human tissues, along with human tissue microarrays, RNA, DNA, and purified proteins	http://www.asterand.com/
Millipore (incorporating Chemicon and Upstate)	Various types of TMAs	http://www.millipore.com/
Folio Biosciences	Human tissue sections and TMAs	http://foliobio.com/
ISU ABXIS	Normal and cancer TMAs, other disease TMAs, frozen sections	http://tissuearray.petagen.com/
Spring Bioscience	Fixed human tissue sections and TMAs	http://www.springbio.com/
ProSci Inc.	TMAs and cell arrays	http://www.prosci-inc.com/
Pantomics	Various types of TMAs	http://www.pantomics.com/

Bioinformatic Approaches for TMA Data Analysis

The Stanford Tissue Microarray Consortium was established to house the Stanford Tissue Microarray Database (TMAD),[13] which stores raw and processed data from TMA experiments along with their corresponding stained tissue images. In addition, TMAD provides methods for data retrieval, grouping of data, analysis and visualization, as well as export to standard formats. Like other databases TMAD data are released to the public at the researcher's discretion or upon publication of a paper. The Recent Publications Portal[14] of the TMAD displays the wide range of research projects using TMAs as well as tutorials, protocols, and bibliography. Browsers may view and search tissues by array block, marker, disorder (NCI ontology), or donor term. In addition, researchers can access web supplements of tissue microarray-based papers that contain a GeneExplorer-based scoring cluster and image database, along with any additional supporting material.

The Yale Cancer Center/Pathology Tissue Microarray Facility[15] was established as a Cancer Center shared resource to provide the Yale community with services related to tissue microarray technologies. This center has a unique interest in automated analysis and quantitative expression analysis in tissues and has developed software (AQUA, a dedicated, automated TMA analysis software) described in a paper in Nature Medicine [23].

Quantification of data from TMAs and image analysis are emerging and controversial areas of research. Over the years pathologists have devised many "semi-quantitative" grading systems. Advances in computing have made morphometric quantitative analysis possible; like cDNA microarrays, the TMA format lends itself to quantitative analysis. However, TMAs also present some special problems that require dedicated readers, or at least dedicated software. An automated analysis protocol must not only be able to select the region of interest, but also normalize it so that the expression level read from any given disk can be compared with other disks. A related problem is that of subcellular localization. Comparisons of nuclear or membranous staining are quite different than total cytoplasmic staining, although there are now a number of automated devices for reading tissue microarrays (BLISS, Chromavision ACIS, Biogenex, Applied Imaging).

TMAs for Toxicological Research:
Opportunities and Challenges

Despite its many advantages, TMA technology has not been embraced by the toxicology research community thus far. Clearly, there are opportunities for toxicologists to use this technology in toxicological immunohistochemistry, especially high-throughput applications. In the next section, a few examples where TMAs have been used by toxicologists are provided.

---

[13] http://tma.stanford.edu/cgi-bin/home.pl

[14] http://tma.stanford.edu/tma_portal/

[15] http://www.tissuearray.org/yale/index.html

Applications in Toxicology

In a recent study carried out in the Republic of Ireland, a team of toxicologists describe the creation of TMAs representative of samples produced from conventional toxicology studies within a large-scale, multi-institutional pan-European project, PredTox, short for Predictive Toxicology, formed part of an EU FP6 Integrated Project, Innovative Medicines for Europe (InnoMed), and aimed to study pre-clinically 16 compounds of known liver and/or kidney toxicity. TMAs were constructed from materials corresponding to the full-face sections of liver and kidney from rats treated with different drug candidates by members of the FP6 consortium. The authors also describe the process of digital slide scanning of kidney and liver sections, in the context of creating an online resource of histopathological data [24].

Conclusions

The driving force behind the development of TMAs was to take the genomics approach already established for DNA and protein microarrays and apply it to the study of clinical tissue samples. TMAs are a significant advance over previous attempts to put multiple samples in a single paraffin block [16, 17]. There are many benefits to using TMAs: one is the ability to screen large numbers of cases in a single immunohistochemical staining run, thereby minimizing day-to-day and batch-to-batch experimental variability in immunohistochemical staining. TMAs also significantly decrease the cost of conducting immunohistochemical studies, and increase the numbers of studies that can be analyzed on small pieces of tissue, by using small cores of tissue rather than cutting sections of every block for each study. DNA microarray technology can be combined with TMA technology to help understand fundamental biological phenomena such as tissue-specific gene expression. This option should be particularly useful for toxicologists. Thus far, in our own laboratory, we have successfully used TMA technology to study the expression of the aquaporin (AQP) water channels and isoforms of Na, K-ATPase in normal human tissues [18–22] and in pathologies of human articular cartilage [25]. A marriage of these two techniques has been applied in the field of oncology to study cancer development and metastasis and identify novel markers that may predict the aggressive behavior of tumors. If properly used, these technologies have the potential for increasing the throughput of biomarker studies in experimental, clinical, and environmental toxicology. TMA technology is likely to have an important role in toxicology laboratories.

Acknowledgments The author wishes to acknowledge the support of Dr. Christopher Moskaluk (University of Virginia), Dr. Stephen Hewitt (TARP Lab, NIH), and Dr. Rachel Airley (University of Huddersfield, UK).

References

1. Heal JR, Roberts GW, Raynes JG, Bhakoo A, Miller AD, et al. Specific interactions between sense and complementary peptides: the basis for the proteomic code. Chembiochem. 2002;3(2-3):136–51.
2. McCulloch A, Bassingthwaighte J, Hunter P, et al. Computational biology of the heart: from structure to function. Prog Biophys Mol Biol. 1998;69(2-3):153–5.
3. Bassingthwaighte JB. Strategies for the physiome project. Ann Biomed Eng. 2000;28(8): 1043–58.
4. Thongboonkerd V. Proteomic analysis of renal diseases: unraveling the pathophysiology and biomarker discovery. Expert Rev Proteomics. 2005;2(3):349–66.
5. Kennedy S. The role of proteomics in toxicology: identification of biomarkers of toxicity by protein expression analysis. Biomarkers. 2002;7(4):269–90.
6. Heller MJ. DNA microarray technology: devices, systems, and applications. Annu Rev Biomed Eng. 2002;4:129–53.
7. Kumble KD. Protein microarrays: new tools for pharmaceutical development. Anal Bioanal Chem. 2003;377(5):812–9.
8. Sanchez-Carbayo M. Antibody arrays: technical considerations and clinical applications in cancer. Clin Chem. 2006;52(9):1651–9.
9. Uttamchandani M, Wang J, Yao SQ. Protein and small molecule microarrays: powerful tools for high-throughput proteomics. Mol Biosyst. 2006;2(1):58–68.
10. Schena M, Shalon D, Davis RW, Brown PO. et al. Quantitative monitoring of gene expression patterns with a complementary DNA microarray. Science. 1995;270(5235):467–70.
11. Bubendorf L, Kolmer, M, Kononen, J, et al. Hormone therapy failure in human prostate cancer: analysis by complementary DNA and tissue microarrays. J Natl Cancer Inst. 1999;91(20):1758–64.
12. Kallioniemi OP. Biochip technologies in cancer research. Ann Med. 2001;33(2):142–7.
13. Kallioniemi OP, Wagner U , Kononen J, Sauter G, et al. Tissue microarray technology for high-throughput molecular profiling of cancer. Hum Mol Genet. 2001;10(7):657–62.
14. Henshall S. Tissue microarrays. J Mammary Gland Biol Neoplasia. 2003;8(3):347–58.
15. Kononen J, Bubendorf L, Kallionimeni A, Bärlund M, et al. Tissue microarrays for high-throughput molecular profiling of tumor specimens. Nat Med. 1998;4(7):844–7.
16. Wan WH, Fortuna MB, Furmanski P. A rapid and efficient method for testing immunohisto-chemical reactivity of monoclonal antibodies against multiple tissue samples simultaneously. J Immunol Methods. 1987;103(1):121–9.
17. Battifora H. The multitumor (sausage) tissue block: novel method for immunohistochemical antibody testing. Lab Invest. 1986;55(2):244–8.
18. Mobasheri A, Moskalukb CA, Marplesc D, Shakibaei, M, et al. Expression of aquaporin 1 (AQP1) in human synovitis. Ann Anat. 2010;192(2):116–21.
19. Floyd RV, Wraya S, Martín-Vasallob P, Mobasheri A, et al. Differential cellular expression of FXYD1 (phospholemman) and FXYD2 (gamma subunit of Na, K-ATPase) in normal human tissues: a study using high density human tissue microarrays. Ann Anat. 2010;192(1):7–16.
20. Mobasheri A, Airley R, Hewitt SM, Marples D, et al. Heterogeneous expression of the aquaporin 1 (AQP1) water channel in tumors of the prostate, breast, ovary, colon and lung: a study using high density multiple human tumor tissue microarrays. Int J Oncol. 2005;26(5):1149–58.
21. Mobasheri A, Marples D. Expression of the AQP-1 water channel in normal human tissues: a semiquantitative study using tissue microarray technology. Am J Physiol Cell Physiol. 2004;286(3):C529–37.
22. Mobasheri A, Marples D, Young IS, Floyd RV, et al. Distribution of the AQP4 water channel in normal human tissues: protein and tissue microarrays reveal expression in several new ana-tomical locations, including the prostate gland and seminal vesicles. Channels (Austin). 2007;1(1):29–38.

23. Camp RL, Chung GG, Rimm DL. Automated subcellular localization and quantification of protein expression in tissue microarrays. Nat Med. 2002;8(11):1323–7.
24. Ryan D, Mulrane L, Rexhepaj E, Gallagher WM, et al. Tissue microarrays and digital image analysis, Methods Mol Biol. 2011;691:97–112.
25. Trujillo E, Gonzalez T, Marín R, Martín-Vasallo P, Marples D, Mobasheri A. et al. Human articular chondrocytes, synoviocytes and synovial microvessels express aquaporin water channels; upregulation of AQP1 in rheumatoid arthritis. Histol Histopathol. 2004;19(2):435–44.

Chapter 3
Importance of Fluorescent In Situ Hybridization in Rodent Tumors

Ana Mendes-da-Silva, Filomena Adega, and Raquel Chaves

Introduction

Cancer is a heterogeneous disease characterized by changes in the genetic and epigenetic cellular background [1]. The morphological and molecular diversity is often observed in tumor cells and emerges as a result of genomic instability. This tumor hallmark is key to the acquisition of mutations essential for the disease progression [2], which may be manifested in various forms in the tumor genome, namely at the level of **chromosomal instability (CIN)** [3]. This type of instability has been considered one of the changes with the greatest impact on tumor progression and is clinically associated with poor prognosis, metastasis, and resistance to chemotherapeutics [4]. Since Boveri's observations of aneuploid karyotypes in cancer cells at the beginning of the twentieth century, the role of CIN in tumorigenesis has been a central issue in cancer biology [5]. Recent studies using sophisticated mouse modeling approaches demonstrated that CIN plays a causative role in a substantial proportion of malignancies [6–8]. CIN can be manifested in terms of numerical chromosomal abnormalities, mainly aneuploidy (whole chromosomal instability, W-CIN), as well as changes in chromosome structure, that can involve equal exchange of material between two chromosome regions (balanced/reciprocal rearrangements) or nonreciprocal chromosomal translocations, jumping translocations, deletions, amplifications, insertions, and/or inversions (structural or

A. Mendes-da-Silva • F. Adega • R. Chaves (✉)
University of Trás-os-Montes and Alto Douro, Department of Genetics and
Biotechnology (DGB), Laboratory of Cytogenomics and Animal Genomics (CAG),
5001-801 Vila Real, Portugal

University of Lisboa, Faculty of Sciences, BioISI– Biosystems & Integrative
Sciences Institute, Campo Grande, Lisboa, Portugal
e-mail: rchaves@utad.pt

© Springer Science+Business Media New York 2016
S.A. Aziz, R. Mehta (eds.), *Technical Aspects of Toxicological
Immunohistochemistry*, DOI 10.1007/978-1-4939-1516-3_3

segmental chromosomal instability, S-CIN) [6–8]. Typically, a structural rearranged chromosome generated from events involving two or more chromosomes or multiple events within a single chromosome is known as a derivative chromosome. Specific regions of the genome can be amplified and the amplified sequences are often present in cancer cells as small fragments called double minutes (dmin's), incorporated into chromosomes in nearly contiguous homogeneously staining regions (hsr's) or interspersed in the genome [8]. It is also common the presence of small chromosomes that appear to be mitotically stable but cannot be classified by conventional banding studies, known as marker chromosomes [9]. The chromosomal rearrangements may alter genes essential for the tumorigenic process, including oncogenes, tumor suppressor, and DNA repair genes by amplifications or deletions, and may be associated with the formation of fusion genes (with oncogenic potential) and transcriptional deregulation as a result of translocations, insertions, or inversions [10].

Cancer genomes are enormously diverse and complex in terms of sequence and structure compared to normal genomes and among themselves [11]. While leukemias, lymphomas, and sarcomas are characterized by the occurrence of recurrent chromosome breakpoints that are type specific, epithelial cancers (e.g., breast, lung, prostate carcinoma) are characterized by displaying more complex genomes, involving both structural and a type-specific pattern of chromosome gains and losses [11–14]. In fact, structural defects in chromosomes are present in a high percentage of both hematological and solid tumors, but in the former, often only a single balanced translocation can drive tumorigenesis [15]. Because of this nonrandom nature of chromosomal abnormalities, several aberrations are diagnostic and prognostic markers, such as BCR/ABL fusion gene present in the Philadelphia chromosome (derivative human chromosome 22) associated with the t(9;22)(q34;q11.2) in chronic myeloid leukemia (CML) [9, 15].

The analysis of specific patterns and the precise determination and identification of recurrent chromosome rearrangements plays a crucial step for understanding and researching the genetic causes for tumor initiation and development, as well as it allows exploration of potential therapeutic targets. Several cancer models have been used and developed to study the carcinogenic process in a **comparative oncogenomic** perspective. In this context, rodents, domestic mouse (*Mus musculus*) and the Norway rat (*Rattus norvegicus*), are particularly prominent [16–19]. The use of rodent cancer models was boosted by the sequencing projects of human [20], mouse [21], and rat genomes [22], which showed a high similarity between rodent and human genetics. Currently, the possibility to access powerful web servers and databases, in which syntenic regions can be easily identified (e.g., http://www.ensembl.org/), smoothes the identification of orthologous genes and pathways involved in tumorigenesis, allowing the association between human and rodent genetics [23]. Despite some of the limitations of rodent cancer models (e.g., [16, 18, 24], most of the important pathways implicated in cancer are driven by orthologous genetic events present in both murine and human tumors [25]. A range of rodent models for human cancer have been described. These include spontaneous models observed in susceptible strains; in models in which tumors are induced by carcinogens, and in transgenic and knockout models where tumors arise as a result of distinct introduced genetic lesions (for more details, see The Cancer Models database http:// cancermodels.nci.nih.gov/ and the eMICE website http://emice.nci.nih.gov/). This latter approach allows the generation of strains carrying conditional oncogenes and tumor suppressor genes, which can closely mimic the sporadic character of human cancers [18, 26, 27]. Several studies demonstrate the importance of these models in studying gene functions (e.g., [28–30]) in identifying novel cancer genes (e.g., [31]) and tumor biomarkers (e.g., [32]), in testing novel therapeutic strategies (e.g., [17, 33]), as well as in the analysis of the molecular and cellular mechanisms underlying susceptibility and tumor onset and progression (e.g., [27, 34, 35]).

The knowledge of the normal chromosomal complement of mouse ($2n=40$; [36]), rat ($2n=42$; [37]) and human ($2n=46$; [38]) was the starting point for comparative analysis of chromosomes in rodent tumors (e.g., [39, 40]) Since Boveri's studies growing enthusiasm in the analysis of chromosomes in tumor genomes developed the discipline of **cancer cytogenetics**. Several improvements in chromosome analysis, mainly in cell culture for metaphase chromosomes preparation and microscope advances (for detail see [41]), led to the development of chromosome banding techniques, termed as **classical cytogenetics**, which enable the study of chromosome numbers and structural integrity at a single cell level [42]. These approaches allow a more precise identification of chromosomes based on their banding pattern (e.g., G-, R-, and C-banding). The discrimination of each chromosome allowed the initial improvement in the characterization of chromosomal abnormalities in rodent tumors (e.g., [43]) and, consequently, the characterization of cancer causing genetic events (e.g., [44]). However, since all mouse chromosomes are acrocentric and are relatively uniform in size, its identification is complicated. Thus, the identification of rearranged mouse chromosomes can be a complex task if assisted only by conventional banding techniques [12]. Similarly to the situation in human solid tumors, the comprehensive identification of chromosomal abnormalities in rodent models of epithelial cancer, for instance, is often difficult due to their complex genomes, with multiple and complex chromosomal changes acquired during tumor progression, and also due to the limitations in cell culture and in preparation of high-quality metaphase spreads [45, 46]. As a consequence, models for leukemia and lymphoma are more thoroughly investigated than solid tumors [46, 47].

The introduction of **molecular cytogenetic** techniques, Fluorescent in situ Hybridization (**FISH**)-based approaches, has helped to overcome some of the limitations of classical cytogenetic techniques, which are still widely used as a standard method for the initial analysis of chromosomes [23, 48]. Importantly, the development of FISH approaches led to an increase of studies both in human and rodent solid tumors [47]. In fact, since its introduction in 1986 by Pinkel et al. [49] and Cremer et al. [50], FISH technology has proved to be a powerful approach and has revolutionized the field of **oncogenomics**. In addition to gene mapping, this approach allows the assessment of copy number changes and the identification and detailed characterization of structural chromosomal abnormalities with high sensitivity and specificity in human and rodent hematological and solid tumors. The main advantage of FISH is the analysis at the single-cell level that is extremely important due to the heterogeneity of primary tumors and metastases [51], which have demonstrated the existence of cell clones within the tumor with different chromosomal content (massive genomic intratumoral heterogeneity) [51, 52].

Fluorescent In Situ Hybridization: A Conceptual Approach

FISH is a method (cf. Fig. 3.1a) used to analyze the presence and location of specific nucleic acid sequences (DNA or RNA) in cytological targets, such as cells, metaphase chromosomes, interphase nuclei, extended chromatin fibers, or transcripts using fluorescence (fluorochromes are chemical compounds that absorb light energy of a specific wavelength and reemits light at a longer wavelength). This is a rapid and sensitive technique based on nucleic acid hybridization between a labeled single-stranded DNA/RNA probe (in Fig. 3.1b are the different types of DNA probes used) and the denatured target sequence. The probe can be directly labeled with fluorochromes (**direct labeling** with nucleotides that have been modified to contain a fluorochrome such as FITC-dUTP, rhodamine-dUTP, etc.) or **labeled indirectly** with modified nucleotides that contain a hapten (e.g., biotin-dUTP digoxigenin-dUTP). Several methods can be used for probe labeling, such as nick translation, random primed labeling, and polymerase chain reaction (PCR) [47]. Before hybridization, both probe and target are denatured to yield single-stranded nucleic acids (not needed for single-stranded probes). Their combination in a hybridization step allows the probe to anneal to its complementary sequence on the target. Later unbound probe molecules or unstable hybrids are removed by stringent washes [53]. When using a probe directly labeled, the fluorescent signals can be immediately seen in a **filter-equipped epifluorescence microscope** and the images digitally captured by appropriate softwares [54]. If the probes are indirectly labeled, it is required an additional detection step consisting in the incubation with fluorescent-conjugated antibodies/anti-hapten (avidin/streptavidin-FITC, anti-digoxigenin-rhodamine, etc.) [47]. In this system, it is possible to simultaneously or sequentially detect as much targets as the number of available microscope fluorescence filters and fluorochromes, as these molecules can use very narrow excitation/emission spectrum wavelengths.

The most widely used targets are **metaphase chromosomes**, providing the possibility to study whole chromosomes or chromosome regions. Its sensitivity in terms of spatial resolution (5–10 Mb) is limited essentially by the degree of chromosome condensation [47, 55]. For metaphase chromosome analysis it is necessary that the sample exhibits mitotic activity per se or that it is prone to cell culturing and propagation. The only way to analyze (fixed) chromosomes is to block cell division in metaphase, the cell cycle stage in which chromosomes are clearly visible as individual entities. This is made by cells' exposure to mitotic blocking agents (e.g., colchicine, colcemid, vinblastine) [45]. But one of the great advantages of FISH technology is the possibility to analyze nondividing cells by the identification of some chromosomal aberrations in **interphase nuclei** (Interphase-**FISH**)

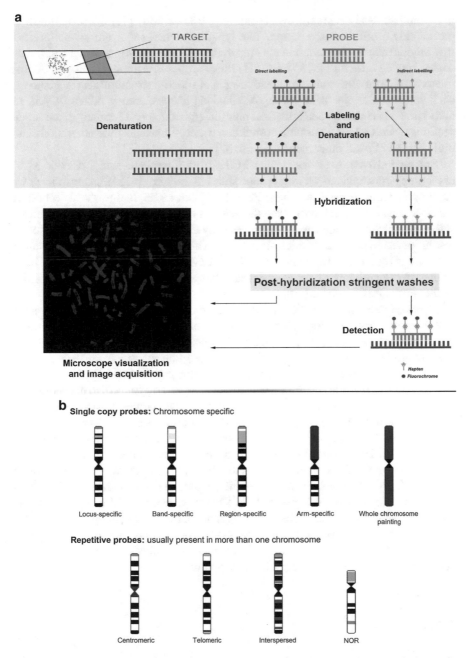

Fig. 3.1 Workflow of fluorescent in situ hybridization (FISH) methodology (**a**) and the different types of probes used for DNA-FISH (**b**)

and simultaneously increase the resolution (50 kb–2 Mb) [47]. This approach is particularly useful in some tumors, like lymphomas and solid tumors that present low mitotic indexes and wherein the preparation of high-quality metaphase chromosomes is often difficult [45, 55]. I-FISH allows a rapid screening of numerical chromosomal abnormalities and the identification of specific structural rearrangements, such as translocations using breakpoint-flanking probes, among others [9, 45]. In addition, I-FISH can be performed on paraffin-embedded and formalin-fixed tissue sections, which allow a possible correlation between chromosome aberrations and biologic and clinical outcomes (e.g., [56, 57]).

Released **chromatin fibers** (Fiber-FISH) from interphase nuclei provide high-resolution target analysis on microscope slides (5–500 kb) [47, 58]. Due to its powerful application in gene mapping, Fiber-FISH can be applied to the detection of chromosomal rearrangements when the analysis of the order of different sequences is the main objective and cannot be accessed by FISH-based approaches with lower resolution, as in the case of micro-inversions. This approach allows accurate sizing of gaps and overlaps between probes [47]. However, the use of Fiber-FISH has been limited to research laboratories rather than clinical ones, because of the complexity of the procedures involved [59].

The need to define even with higher-resolution chromosomal rearrangements led to the development of cytogenetic analysis approaches that are beyond the cytological targets. This was achieved by the introduction of **DNA microarrays** in the cytogenetic analysis [60]. Generally, microarrays are constructed from various-sized genomic targets ranging from bacterial artificial chromosome (BACs; 80–200 kb in size) clones to synthetic oligonucleotides (25–85 bp in length), which are spotted on a glass surface on the array [61, 62]. Once the genomic targets are spotted strategically in the array, where genomic position of each clone is exactly known, it is possible to identify specific chromosomal abnormalities such as amplifications and deletions [62]. The resolution of the array depends on the genomic coverage, which is restricted by the density and size of the clones [47, 62]. Array platforms can be designed for whole genomes or for specific genome regions, such as specific chromosomes or chromosomal segments [63]. Currently, several platforms are commercially available for several species, namely human, mouse, rat, among many others (e.g., http://www.microarrays.com/). Section "Microarrays Based Cytogenetics" will address the types of microarrays for cytogenetic analysis and its importance in detecting chromosomal abnormalities in rodent tumors.

DNA Probes

Typically, the probe consists of a cloned nucleic acid (mainly a DNA) sequence. In addition, probes of oligonucleotides and synthetic nucleic acids, peptide nucleic acids (PNA) [28, 64], and Locked Nucleic Acids (LNA) [65, 66] can be designed for improvement of both sensitivity and resolution. According to the target sequence, the probes can be subdivided into two major classes: single copy and repetitive probes (Fig. 3.1b). The first correspond to probes that are specific for unique sequences, which are typically present in euchromatin and in specific *loci*. This class includes probes for a particular *locus* (genes), band, region, or even chromosome arm or whole chromosome. The second class of probes targets sequences that are naturally repeated in the genomes (repetitive) and the majority are located in heterochromatin. These regions are present in centromeres, the primary constrictions of chromosomes, and in telomeres, but can also be found interstitially along the entire chromosome. The majority of these sequences are common to a number of chromosomes from the complement (some—pan—are common to all the chromosomes). Nevertheless, there are also chromosome-specific repetitive probes, as it is the case of human chromosomes. In addition, probes for ribosomal DNA genes (NOR probes), which are tandemly arranged, are also grouped in this class.

Single-Copy Probes

The sequencing of genomes from several species, including human, mouse, and rat, has enabled the availability of genomic DNA sequences (especially coding genes) cloned in vectors, such as bacterial artificial chromosomes (BACs) or P1-derived artificial chromosomes (PACs), which assists the generation of *locus*-**specific probes** (LSPs) [12, 53]. Considering the spatial resolution of the cytological targets, the use of these vectors permits the detection of specific genomic regions, such as protein-coding genes that typically comprise about 100 kb [29], allowing performing gene mapping. In FISH experiments using LSPs (BAC clones), hybridization occurs only between the insert in the vector (sequence of interest) and the complementary target sequence. However, since the total BAC vector is labeled, even the regions not involved in the hybridization are also detected, and targets as small as genes can be identified.

In addition to gene mapping, LSPs can be used for assessing copy number changes involving specific chromosome regions/genes, such as deletions, duplications, and dmin's that are common in solid tumors [67, 68]. These probes are also a valuable resource in the validation and more detailed characterization of inter- and intrachromosomal rearrangements, such as translocations and inversions [23]. LSP probes can be designed for a specific gene, a translocation breakpoint, or another DNA sequence [12, 53]. Additionally to metaphase spreads' analysis, these probes can also be used in interphase nuclei, which allow a rapid screening of chromosomal abnormalities that are often tumor specific, such as the *BCR/ABL* fusion gene associated with the translocation t(9;22)(q34;q11.2) in chronic myeloid leukemia (CML) and the *PML/RARA* fusion gene associated with t(15;17)(q22;q12) in acute promyelocytic leukemia (APL). LSPs spanning the breakpoint region can be applied in a dual-color FISH experiment for two different chromosome regions/genes involved in a reciprocal translocation, and the translocation can thus be identified by color overlap in interphase nuclei [9]. Currently, LSPs are commercially available for the referred and other specific chromosome translocations followed by chimeric gene fusions that are common in hematological disorders [9].

Probes for larger genomic regions are also widely available; entire genomes (described in more detail in section "Multicolor Banding") are probed for specific chromosomes (known as **whole chromosome painting** (WCP) probes) [69, 70], **chromosome arms** [71], **bands, or regions** [72]. These specific probes are typically obtained by procedures that allow isolation of specific chromosomes or chromosome segments. To date, the most successful and almost universally used approach for preparing probes of entire chromosomes is **fluorescent activated cell (chromosome) sorting** (FACS), a method designed to physically separate chromosomes based on their size and AT/GC composition [73]. Then, the DNA correspondent to a specific chromosome is subjected to a primary degenerated oligonucleotide primed PCR (DOP-PCR) with a universal primer (e.g., 6MW) [70]. The primary PCR products are further amplified in a second DOP-PCR, thus providing the template for labeling [74]. In the case of mouse chromosomes, their similar size makes the isolation of each chromosome by FACS difficult [75]. However, the use of specific mouse strains has overcome these difficulties [46]. Another approach, **laser microdissection**, provides an excellent opportunity to isolate chromosomes or chromosome segments (a chromosome arm, band, or region) [76]. This technology employs a focused laser beam to extract a defined microscopic area in metaphase chromosomes, which were previously transferred onto a microscope slide covered with a polyethylene–naphthalate supporting membrane. Probes are generated by laser microdissection by catapulting from 8 to 15 copies of the chromosome or segment of interest from the slide into a microtube [77]. This material is then amplified by the DOP-PCR-based method referred above and direct/indirectly labeled [77, 78]. Other methods like nick translation or random primed labeling are also suitable for labeling these probes [78]. It is important to notice that these probes are not "unique" in their sequence. Actually, it may contain disperse repetitive elements or, in case of WCPs, also centromeric repeats, which can result in unwanted nonspecific hybridization all over the genome. This problem is overcome by in situ suppression hybridization [69], where repetitive sequences are suppressed by addition of an excess of unlabeled DNA fractions enriched in repetitive sequences (e.g., cot-1 DNA) to the hybridization mixture [12, 69]. Currently, in addition to human, mouse, and rat, chromosome-specific paints are commercially available for a vast number of vertebrate species.

The main application of WCPs is the detection of translocations involving nonhomologous chromosomes [12], being the detection of intrachromosomal changes, like deletions, duplications, or inversions limited with these probes (Fig. 3.2). However, the combination of these with other probes is an excellent resource for the cytogenetic characterization of rodent tumors (e.g., [23, 48, 67]).

Works carried out by our group reflect the applicability of WCPs and LSPs in the cytogenetic characterization of rodent tumors. Louzada et al. [23] performed the first cytogenetic characterization of HH-16 cl.2/1 rat mammary tumor cell line using a multi-cytogenetic approach, which included G-banding, chromosome painting using rat and mouse probes, and BAC/PAC clones' hybridization [23]. G-banding permitted a global analysis of the numerical and structural chromosomal abnormalities, and WCPs and BAC/PAC clones (LSPs) allowed the characterization of the

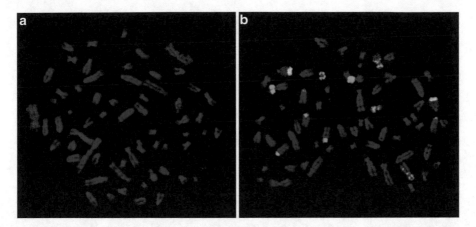

Fig. 3.2 Representative images of WCP with painting probes from *Rattus norvegicus* and *Mus musculus* in metaphase preparations from the rat tumor cell line CLS-ACI-1: (**a**) rat painting probe for chromosome RNO6 (labeled with digoxigenin and detected with anti-dig-rhodamine, *red* signal); (**b**) mouse chromosomes MMU5 (labeled with biotin and detected with avidin-FITC, *green* signals) and MMU11 (labeled with digoxigenin and detected with anti-dig-rhodamine, *red* signals)

clonal evolution of this cell line by the accurate detection of complex chromosome rearrangements and breakpoint regions. This analysis enabled the identification and characterization of different cell subclones, which allowed inferring ancestral rearrangements as well as the reconstruction of the clonal evolution that occurred during this tumor progression. *In silico* analysis of the breakpoint regions detected in the cell line demonstrated that almost all contain genes in the rat genome for which the human homolog has been associated with breast cancer. Among these, *Erbb2* and *Mycn* oncogenes were selected by the authors for further FISH analysis with specific BAC clones, which revealed their amplification (one of the mechanism for oncogene activation) [23]. This study allowed a deeper characterization of this cell line, and importantly, it demonstrated its potential as a model for breast cancer research.

Finally, probes generated from rearranged chromosomes can be applied for reverse chromosome painting, called **reverse-FISH** [79]. In this approach, WCP probes generated from flow-sorted or microdissected rearranged chromosomes are hybridized onto normal metaphase spreads and the content and breakpoints of that rearranged (derivative) chromosome can be revealed [47, 79]. In a mouse model of plasmacytoma (BALB/c mouse plasmacytoma, MOPC 315), typically harboring a balanced t(12;15) chromosomal translocation deregulating the expression of the proto-oncogene *c-myc* (by the result of the *IgH/Myc* fusion gene), two quimeric marker chromosomes were analyzed by reverse-FISH, which allowed the characterization of their composite nature and the exhibition of the t(12;15) breakpoint region in both of them, probably by the occurrence of a jumping translocation involving this region [80].

Each type of probe has a specific ability and provides very specific answers about chromosomal abnormalities [45]. The choice of the right probe or of the simultaneous use of a pool of probes in a multicolor approach, as well as the election of the most suitable target, depends on the purpose in question, the resolution needed, and the detail of information required.

Whole-Genome Analysis in a Single Experiment with FISH-Based Technologies

The development of metaphase chromosome analysis, as well as a vastly increased number and combination of probes, in association with probe labeling developments and significant advances in optical microscopy, allows nowadays a detailed and thorough analysis of tumor genomes. Currently, the screening of entire tumor genomes for chromosomal abnormalities is possible in a single experiment without prior knowledge of the chromosomal segments involved.

Multiplex-FISH and Spectral Karyotyping Analysis

The use of differently labeled WCPs combined in a single experiment led to the development of two main technologies that allow to delineate each individual chromosome of the complement in a single experiment: multiplex-FISH (**M-FISH**) using different fluorochrome-specific filters and a combinatorial labeling algorithm [81], spectral karyotyping (**SKY**) similar to M-FISH but using an interferometer-bascd spectral imaging [82], and combined binary ratio labeling (**COBRA**) that use simultaneously combinatorial and ratio labeling for probe discrimination [83]. By using computer-generated false colors, each chromosome is imaged with different colors, and a multicolor karyotype is obtained.

Since all chromosomes are analyzed simultaneously, no *prior* knowledge of the affected chromosome(s) is required and complex rearrangements in tumor cells can be rapidly identified, which is a great advantage in the analysis of tumor karyotypes that can be often highly complex. In a single procedure, each of these approaches is capable of detecting numerical and structural interchromosomal abnormalities involving large regions, such as interchromosomal translocations, dmin's, and hsr's [12, 84]. In fact, next after numerical chromosomal alterations and translocations in tumor cells, formation of unidentifiable marker chromosomes of multiple chromosomal origins is often common [85] and SKY has assisted greatly in the identification of the origin of such marker chromosomes [86]. The identification of aberrant chromosomes in rodent tumor models was greatly simplified by the introduction of M-FISH/SKY analysis on mouse [82] and rat chromosomes [87], and the current commercial availability of WCP probes (Applied Spectral Imaging, Inc). A wide variety of rodent models of human cancer, including hematological malignancies and solid tumors, were cytogenetically characterized and analyzed with a M-FISH/ SKY approach (e.g., [29, 80, 82, 86, 88, 89]). With this approach, Coleman et al. [80] analyzed ten established BALB/c plasmacytomas in which the t(12;15) harboring the *IgH/Myc* fusion had been previously detected by G banding. But the sequential SKY analysis revealed that secondary chromosomal rearrangements were contributing to the tumor progression in BALB/c mice [80, 90, 91]. Recently, Falck and coworkers [86] used SKY to identify cancer-related aberrations in a well-characterized experimental model for spontaneous endometrial carcinoma (EC) in the BDII rat tumor model. The analysis of a set of 21 BDII rat EC revealed nonrandom numerical and structural chromosome alterations, including gain of chromosome 4, loss of chromosome 15, and structural rearrangements in chromosomes 3, 6, 10, 11, 12, and 20. Importantly, the nonrandom nature of these events suggests their potential contribution in the development of BDII EC [86]. Having the syntenic segments between rat and human as a guide and the analysis of specific genes located in chromosomes and regions involved in structural abnormalities in both rat and human EC, the discovery of potential genes influencing EC onset and progression is possible.

However, taking into account that the probes for SKY/M-FISH analysis are usually WCPs from the same species as the target, detection of rearrangements within the same chromosome like intrachromosomal translocations, duplications, deletions, and inversions is almost impossible (involving <3 Mb of sequence) [47, 55]. The same happens for very small markers or dmin's, which cannot usually be unambiguously classified [55]. In these cases the limitations of SKY/M-FISH can be supplemented with other FISH-based approaches [90, 91] that will be addressed below.

Comparative Genomic Hybridization Analyses

Comparative genomic hybridization (CGH) technology was developed to overcome the difficulties in preparing high-quality metaphase spreads from solid tumors [92]. CGH is a potent and reliable technique that, in a single hybridization experiment, identifies and maps DNA copy number changes (gains and losses) on chromosomes (conventional CGH, i.e., cCGH), providing detailed information about tumor genomes [93]. Particularly frequent in solid tumors, copy number aberrations are expected to contribute to tumor evolution by inducing alterations in gene expression. Once no initial knowledge of the chromosome imbalance in the genome is required, CGH is used as a "discovery tool" [47]. This approach is based on the hybridization of differentially labeled tumor (test) and reference DNA from the standard counterpart (e.g., in red and green, respectively) on normal metaphase chromosomes under in situ suppression hybridization [92, 94]. The probes compete for complementary hybridization sites, which results in differences in fluorescence intensities along the chromosomes that are a consequence of changes in the tumor sample. In general, an even combination of red and green fluorescence is observed in a specific chromosome region if DNA is present in identical copy numbers in both reference and tumor genome [47]. The presence of losses or gains results in a shift to green or red staining, respectively. After image capturing and karyotyping, the fluorescence ratios are quantified using digital image analysis and a copy number change profile of tumor genome along the chromosome length is produced [54, 55]. Quantification is thus a crucial step for a precise CGH analysis. In terms of spatial resolution, its sensitivity is mainly limited by the degree of chromosome condensation on the reference metaphase chromosomes. However, it is also influenced by copy number changes. In fact, despite the resolution limit in metaphase chromosomes (5–10 Mb), amplifications in small regions (minimum 50 kb) can be detected if the copy number change is high [55].

In contrast with other FISH-based technologies, CGH has the enormous advantage of not needing tumor cell culturing and metaphase chromosomes for screening. DNA can be extracted from fresh or frozen tumor tissue or even from formalin-fixed paraffin-embedded tissues [57, 95]. These procedures involve the extraction of DNA from a cell population; however, advances in CGH have enabled its application to the analysis of single cells, allowing the analysis of clonal heterogeneity [96, 97].

CGH has been applied to a variety of rodent models, from solid (e.g., [93]) to hematological tumors (e.g., [98]), and has been a valuable tool for genome-scale analysis. Kogan and coauthors [98], through CGH analysis of murine APL, identified recurrent gain of chromosomes 7, 8, 10, and 15 and recurrent loss of chromosome 2, which, based on orthologous regions, correspond to karyotype abnormalities in human APL. The analysis of chromosome copy number abnormalities in murine APLs may facilitate the identification of genetic changes that cooperate with the fusion protein PML/RARA to cause acute leukemia [98], which can improve the diagnosis, therapy, and outcome of this pathology.

A detailed and precise analysis of human and rodent tumor models allowed the determination of a specific pattern of chromosomal gains and losses that is a landmark of a specific tumor (e.g., [25, 93, 99]). In fact, in most studies, a concordance between the rodent and the human patterns for specific tumors was reported (e.g., [25, 29, 30, 89, 93]). As an example, Christian et al. [93] used the CGH approach to analyze mammary carcinomas in rats that were induced by 2-Amino-1-methyl-6--phenylimidazol[4,5-*b*]pyridine (PhIP), the most prevalent of the known carcinogenic heterocyclic amines in the human diet that was implicated in human cancers as breast cancer [100]. The analysis of rat carcinomas' imbalanced regions seems to represent a signature pattern for PhIP-induced carcinomas. Moreover, these regions are located on orthologous regions of human chromosomes and deletions on these regions are reported as involved in human breast cancer [99]. In addition to specific patterns analysis, several CGH studies in rodent models focus on the study of specific genes that play a known and important role in carcinogenesis (e.g., [98]). The analysis of genomic imbalances in mammary tumors of BRCA1 conditional mutant mice, with a similar pattern to human breast carcinomas [29, 30, 89], reveals that chromosomal regions involved in genomic imbalances are orthologous to human, and harbor tumor-related genes like *Erbb2*, *p53*, or *c-Myc*. The CGH analysis performed by Weaver and colleagues [89] revealed a gain and loss, respectively, in the distal and proximal regions of chromosome 11, syntenic to human chromosomes 17q11-qter and 17p, which comprise *Erbb2* oncogene and the tumor suppressor gene *p53*. Furthermore, a whole or partial gain of chromosome 15 and consequently of *c-Myc* oncogene was also observed. Recently, Heiliger et al. [31] analyzed copy number changes in 30 murine tumors from aTrk-T1 transgenic mouse model of thyroid neoplasia. The copy number analysis of murine thyroid tumors revealed novel gene alterations that are also prevalent in the human counterpart [31]. The comparison of genomic abnormalities in tumor genomes in rodent models and human cancer allows the discovery of novel candidate genes with a crucial function in the tumoral process, as well as a better understanding of the processes underlying carcinogenesis.

The applicability and advantages of CGH genome analysis are clear. However, this technology has some limitations. Firstly, no information is provided about the structural arrangements (e.g., duplications, isochromosome formations, dmin's, and hsr's) involved in the gains and losses. Secondly, balanced rearrangements such as translocations and inversions, in which no copy number is changed, cannot be detected. Thirdly, it is not possible to discriminate a whole genome copy number change such as it happens in ploidy [12, 47, 101].

Microarrays Based Cytogenetics

The advent of microarrays provided a description of chromosome abnormalities at a resolution that exceeds the microscopic analysis [47]. The array technology enhanced the cytogenetic analysis with the development of the **array CGH** (aCGH) [60, 94, 102, 103]. In this approach, the tumor and normal genome are differentially labeled and co-hybridized to a microarray (Fig. 3.3 with cCGH and aCGH methodologies). Then, the array is scanned and the relative fluorescence intensities are calculated for each known mapped clone/sequence that is spotted in the array. DNA copy number changes are analyzed by the resulting intensity ratio [75]. Due to the resolution of the analysis that depends on the density and size of the clones on the array [61, 103], it is possible to detect changes in specific *loci*. The highest resolution for array CGH can be provided by high-density oligonucleotide arrays, which can detect very small alterations (100 kb) [62, 103]. Currently, another type of array is available for copy number changes assessment, which can provide resolution at a single nucleotide level, single nucleotide polymorphism-based arrays (**SNP arrays**) [47, 62, 104]. In addition to copy number alterations, this platform allows the identification of loss of heterozygosity (LOH) or copy number neutral loss (CNNL) that cannot be detected using CGH-based arrays. SNP arrays allow the study of the contribution of LOH/CNNL to tumor onset and progression (for details see [62]).

In addition to human diagnosis and research, the whole-genome analysis with aCGH has been performed in tumor studies in mouse (e.g., [105–111]) and rat models [112]. It is noteworthy that Hackett et al. [106] analyzed 40 mouse neuroblastoma (initiated by the *TH-Mycn* transgene) using an array comprised of 1056 BAC clones providing 5 Mb average resolution across the genome, and sub-Mb resolution in regions identified in previous genetic screenings [105]. Additionally, specific

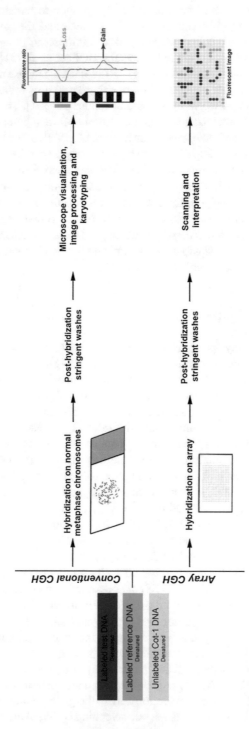

Fig. 3.3 Schematic diagram of comparative genomic hybridization (CGH) methodologies, namely conventional CGH and array CGH

alterations detected by aCGH were validated by FISH with LSPs (BAC clones). Globally, the results obtained by Hackett et al. [105] demonstrated conservation of many genetic changes in mouse and human neuroblastoma. Recently, the same array platform was used by To et al. [110] to analyze DNA copy number alterations across the genomes of lung tumors in the $Kras^{LA2}$ model of lung cancer on the inbred FVB/N mouse strain. This analysis showed a significant association between progressive increase of genomic instability and tumor size. Moreover, on smallest tumors the amplification of $Kras$ was the only detectable change, which can suggest that this may be the earliest genetic event during lung tumor development in the $Kras^{LA2}$ model [110]. Currently, aCGH platforms with variable resolutions are commercially available for mouse and rat genomes (Agilent Technologies). The online availability of CGH data for a wide range of human tumors (Progenetix database http://www.progenetix.org/) in addition to the mapping of conserved syntenies between the human, mouse, and rat genomes (e.g., ensembl database, http://www. ensembl.org/) allows an effective comparison between rodent and human tumors.

Array CGH is a powerful approach for detecting copy number changes with high resolution. However, like conventional CGH, chromosomal abnormalities that do not involve copy number changes, such as inversions or balanced translocations, cannot be accessed by aCGH. In 2003, Fiegler et al. [113] combined the reverse chromosome painting with microarray technology, in a process termed **array painting,** a modification of the array CGH method. In this approach, probes obtained from flow sorted (or microdissected) derivative chromosomes are amplified and the amplified products are differentially fluorescently labeled and co-hybridized onto a microarray [113, 114]. Array painting allows the rapid and efficient high resolution mapping of chromosome translocation breakpoints restricted only by the density of clones on the array. In the last decade, several improvements in the resolution of microarrays for array painting applications were performed [114, 115]. In addition to balanced translocations, array painting can be used in the identification of rearrangements within the derivative chromosomes, such as inversions at the breakpoint region, deletions, and translocations or insertions of extra material from other chromosomes [47, 114].

Multicolor Banding

As referred previously, WCPs are powerless in detecting the precise location (mapping) of the breakpoints preceding a certain chromosome rearrangement or in the identification of an exact intrachromosomal rearrangement. To overcome these limitations, other approaches have been developed [116–118]. A set of subregional chromosome probes or partial chromosome painting (PCP) probes can serve as a base for the so-called **multicolor banding** (MCB/mBAND) [116, 117]. Generally, these subregional segments for MCB probe mixes are generated by microdissection [76, 116]. In some cases, FISH with WCPs may precede microdissection for accuracy in the identification of mouse chromosomes (known as FISH-microdissection) [119].

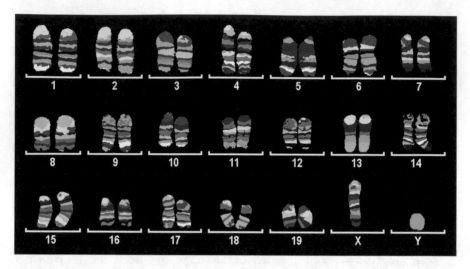

Fig. 3.4 Combined image of multicolor banding (MCB) for all the 19 mouse autosomes and the X chromosome (adapted from [119])

The MCB approach was established for all murine chromosomes (Fig. 3.4) by Thomas Liehr group [90, 91, 119–121]. MCB probes can unambiguously determine peri- and paracentric inversions and map breakpoints, but with metaphase chromosomes resolution. This approach was used for the characterization of rodent tumor cell lines [90, 91, 121, 122]. The first mBAND study for murine chromosomes is reported by Benedek et al. [122]. In this work, MCB probes were developed for chromosome 11, to further characterize the nature of the intrachromosomal rearrangements on this chromosome in BALB/c PreB ABL/MYC lymphocytes. With this approach, chromosomal rearrangements such as simple duplications, inverted duplications, deletions, translocations, and insertions were detected. Later, Karst et al. [90, 91] studied WMP2 cell line using SKY combined with five mouse-specific MCB probes for chromosomes X, 3, 4, 6, and 18, allowing the detection of five previously unrecognizable (by SKY) intrachromosomal rearrangements with nine breakpoints. This study demonstrated that cryptic abnormalities could easily be detected by MCB. More importantly, this effective cytogenetic characterization allows establishing a genotype/phenotype correlation. The continuous improvements in murine MCB probes allowed recently the first genome-wide high-resolution characterization of the extensively used murine tumor cell line NIH 3T3 using all 21 chromosome-specific murine MCB probe sets [121]. In addition to karyotype delineation, this approach can be used for detecting chromosomal imbalances such as the CGH analysis [117, 121]. The gains and losses of NIH 3T3 were aligned with the orthologous regions in the human counterpart in 18 human carcinomas, and a high rate of concordance was observed in the tumors originated from ectoderm (e.g., breast carcinoma, neuroblastoma, and glioma) [121].

In addition to the MCB approach, a YAC/BAC-based chromosome bar code can also work as a FISH-banding method. This has been done for mouse chromosomes

[123, 124]. But there is a clever form of getting a multicolor banding of any genome, at least in theory, by using WCP probes from a different species than the one from the target in analysis (**cross-species/ZOO-FISH**) [125, 126]. This is especially effective when the two species involved are karyotypically very different (due to the course of evolution), but with a certain degree of phylogenetic proximity (due to constraints of the methodology, namely in the time required to produce a stable probe/target hybrid). Comparative genomics play a key role in this issue as it allows choosing the ideal genome to generate the probes. Of course this may not be feasible when working in a strict clinical ground, but the fundamental research has assisted in this issue and today there are already laboratories applying this technology not only in human but also in rodents and other species' cancer cytogenetics and providing this technology to the public in general.

The first and the best known case is the use of gibbon (e.g., *Hylobates concolor*) WCPs to delineate up to 90 different chromosome segments in the human karyotype [125, 126]. But also in rodents this has been done, as it is the case of Buwe et al. [87] for rat chromosomes and later human, mouse, or rat probes that have been used to cytogenetically characterize complex rodent tumor genomes (e.g., [23]). Recently, our group, using this methodology, was able to precise the exact rat chromosome segment involved in a clonal rearrangement of the commercial rat tumor cell line HH-16 cl.2/1 with a mouse WCP probe [23]. The choice of the probe to be used was based on the syntenies between rat and mouse assessed by the ensembl database (http://www.ensembl.org/; assembly RGSC 3.4).

Analysis of Centromeres and Telomeres in Tumors

The centromere and the telomere are essential components for the stability and integrity of genomes and for the faithful transmission of the genetic information from one cell generation to the other (for a review, see [127, 128]). Aneuploidy and CIN are hallmarks of the majority of tumors [3] studied so far and can be driven by centromere and/or telomere dysfunction resulting from eroded or unprotected telomere structures [64]. FISH experiments with centromere and telomere probes are not only used in human tumors' diagnosis and research, but also in rodent tumors [28, 64].

Telomeric sequences were highly conserved across evolution. By contrast, centromeric satellite DNA families are extremely diverse, as these repetitive sequences may evolve rapidly. This fact was used to generate chromosome-specific centromeric probes for several species, including human and rat [129, 130]. So far, in mouse this wasn't possible to achieve due to the great centromeric homogeneity that the chromosomes of this species seem to exhibit.

Probes derived from centromeric repeats provide a valuable tool for the identification of aneuploidies and specific rearrangements involving the centromere region, such as marker chromosomes and centromeric fusions. Additionally to G-banding, C-banding is sometimes used as an initial screening of chromosomal instability. By revealing constitutive heterochromatin, this classical cytogenetic technique allows identifying the molecular nature of a particular derivative chromosome or chromosomal segments. Sequential FISH with centromeric/telomeric probes can be performed in an attempt to further investigate the involved region.

Dual-color experiments with centromeric and telomeric probes can be performed for a more detailed analysis of specific rearrangements involving these regions, such as pericentric inversions and tandem and Robertsonian (Rb) translocations [28, 64]. Rb translocations involve the abnormal joining of two acrocentric nonhomologous chromosomes (usually), which typically fuse at the centromeric region. This rearrangement has been observed in hematologic and solid tumors of many species, including human and murine ones [28, 64]. Furthermore, telomeric probes also allow the study of the length of telomeres and, consequently, evaluate genome stability (integrity) in tumor progression [28, 64]. Several are the studies reporting the relation between telomere shortening and tumor progression [64, 131, 132]. The study reported by Silva et al. [64] is a great example of the applicability of centromeric and telomeric probes in rodent tumor analysis. In this study, alterations in the primary constrictions of chromosomes that accompany melanocyte transformation in a murine melanoma model were characterized with centromeric and telomeric PNA probes after an initial screening of chromosomal abnormalities by SKY. A great centromeric instability was detected in this model, manifested by a number of centromeric fragments or dmin's and Rb translocations. This last abnormality was only present in the metastatic population, but all the analyzed metaphases exhibited the nonrandom Rb (8;12) translocation, what may indicate a selective proliferation advantage of cells carrying this Rb fusion during the acquisition of metastatic potential [64]. The use of telomeric probes revealed the gradual decreasing of telomere length in the melanoma progression, achieved by **quantitative FISH** (Q-FISH). In this approach, telomere PNA probes are hybridized on 3D interphase nuclei, and the length of telomeres is measured by specialized software based on fluorescence intensities of individual chromosome ends [28, 64, 133].

The nuclear organization of mouse centromeres and telomeres in primary, immortalized, and tumor murine cells has been performed by 3D-FISH with centromere and telomere PNA probes in mouse nuclei [28, 131, 134]. Significant differences in the nuclear distribution of centromeres and telomeres were observed during tumor progression [28, 134].

Combining Approaches: An Improvement in Cytogenetic Characterization

Given the limitations and advantages of each cytogenetic approach in the precise identification and characterization of the etiological chromosomal abnormalities in cancer, the combination of techniques that complement each other allows a superior enhancement in the characterization of tumor karyotypes (genomes) characterization. The choice of the specific FISH probe or FISH approach is based on the aim of the study, and in particular, of the required information or resolution on the rearrangements involved.

As an example, in addition to a CGH analysis performed by Kogan et al. [98] in a mouse model of APL (cited above), a SKY analysis of the same model by Le beau et al. [135] allowed a more detailed characterization of abnormal clones by the reconstruction of the clonal evolution and the patterns of the abnormalities found. With this combinatorial FISH approach, it was possible to determine that secondary mutations cooperate with the PML/RARA fusion protein and contribute to the development of APL in both murine and human tumors [135, 136]. In a solid tumor, a dual analysis with CGH and SKY on the early passages of mouse lung adenocarcinoma cell lines allowed the identification of target regions with putative susceptibility genes involved in mouse lung adenocarcinoma [34]. If more detailed information about specific rearrangements is required (wanted), mBAND [90, 91] or LSPs [23] can be used. Microarrays' platforms or aCGH have also been used in combination with SKY, or other FISH approaches, for resolution improvement of chromosome imbalances and rearrangements [25].

RNA Analysis in Single Cells

Although DNA analysis is important, expression analysis of particular DNA sequences represents a crucial step in cancer studies. Determination of the spatial and temporal expression patterns of a gene, which can be translated or not, as well as the cell location of its expressed product, is essential for unveiling its biological function. RNA is a powerful marker with several features and cellular processes, including identification of specific cell types, different developmental stages of cell differentiation, and abnormal RNA expression that often follows tumor growth [137].

One of the limitations of most methods for expression analysis is the use of RNA obtained from a starting cell population. Cells in different cell cycle phases or developmental stages have distinct expression profiles that are hidden when using a cell population. Thus, the variability between different cells cannot be studied by analyzing a cell population, which is not physiologically homogeneous. In the last years, **RNA-FISH** (Fig. 3.5) has emerged as a powerful tool for expression analysis in the context of single cells, maintaining the cellular morphology and spatial information [23]. In combination with laser scanning confocal microscopy (3D FISH), RNA-FISH allows detailed analysis of gene expression, including low abundant RNAs, at a single-cell resolution [23], which is extremely important in the context of tumor studies. Specifically, solid tumors consist of a compendium of many distinct cell types: cancer cells, cancer stem cells, cancer-associated fibroblasts, endothelial cells, pericytes, and immune inflammatory cells, besides being known that microenvironment influences gene expression in those cells [2].

RNA-FISH allows detection and localization of specific RNA in a biological context, fixed cells or tissues, and the assessment of the distribution patterns at the

Fig. 3.5 Expression analysis of *Erbb2* oncogene by RNA-FISH (*green*) and ribosomal 18S (*red*) used as reference, in the rat tumor cell line HH-16.cl.4

organismal, cellular, and subcellular levels [138]. One of the crucial steps for RNA-FISH is the best choice of the probe that is suitable for the detection of a specific RNA target. In addition, the probes should have the adequate size to penetrate the cell membranes and tissues, once the cells morphology should be preserved. There are three major classes of probes for RNA-FISH: DNA probes (cDNA or gDNA), RNA probes (riboprobes), and synthetic oligonucleotides. Each of these classes has advantages and limitations; however, the advent of synthetic oligonucleotides and their associated advantages (for details, see [139]) have made them the probe type commonly used for the detection of nuclear and cytoplasmic transcripts. The increase of the oligo probes specificity can be achieved by using modified nucleic acids (PNA and LNA probes), as well as signal amplification systems to increase the sensitivity [140]. Specifically, LNA probes have been widely used in the detection of small transcripts, such as small noncoding RNAs (ncRNAs) [65, 66].

Several studies reporting the success of RNA-FISH as a semi-quantitative approach (e.g., [23, 32]) have been used to analyze *Erbb2* and *Mycn* genes expression in the HH-16 cl.2/1 and HH-16.cl.4 rat mammary cell lines by real-time reverse transcription quantitative PCR (RT-qPCR) complemented and validated with an RNA-FISH analysis. In fact, RNA FISH data strongly supported the RT-qPCR expression results demonstrating to be an excellent technology when applied either alone or together with other technique [23], from the analysis of expression patterns of crucial genes on tumor evolution [23] to the discovery and validation of biomarkers of cancer [32, 65, 141]. One of the major areas of application is the analysis of ncRNAs, like miRNAs, especially regarding the establishment of its biological function in a normal or pathological context [32, 65, 142]. As an example, Liu et al. [65] analyzed the MicroRNA-31 (miR-31) expression profile in paired normal-malignant lung tissues from mouse and humans by RNA-FISH with an LNA probe. This miRNA acts as an oncogenic miRNA (oncomir) in lung cancer by targeting specific tumor suppressors for repression. In this study, an overexpression of miR-31 was detected within murine and human malignant lung tissues [65].

For a more accurate analysis, different approaches to cell imaging can be performed. RNA-FISH and DNA-FISH can be performed in association with immuno-fluorescence (**immuno-FISH**) allowing the determination of the spatial association of specific genes and/or transcripts with a particular cellular compartment, structure, or proteins [139]. This approach provides information about the relationship among the transcription level, the cellular location, and the protein expression in individual cells. Mouse studies have used this strategy for improving the expression profiles and biological function of specific RNAs [32, 142, 143]. As an example, Deng et al. [32] analyzed the regulation and function of telomeric repeat-containing RNA (TERRA) with a PNA probe in normal and cancer cell development in human cancer biopsies and in the mouse model for medulloblastoma. TERRA is a nuclear heterogeneous length noncoding RNA, transcribed through the terminal telomere repeats of eukaryotic chromosomes [144], and can be developmentally regulated [32]. The analysis performed with DNA-FISH, RNA-FISH, and immunofluorescence demonstrated that TERRA accumulates and forms TERRA *foci* (TERFs) in highly proliferating progenitor neurons and cancer cells. According to all the results, Deng et al. [32] considered TERFs a powerful biomarker for telomere dysfunction in cancer.

The description of transcriptional profiles in combination with genomic data from rodent tumors not only allows a better understanding of the biological processes that shape tumor genome but also validates animal models for the study and understanding of human tumors.

Concluding Remarks

Cancer is a genomic disease characterized by a great chromosome instability where the identification and characterization of chromosomal alterations allow the detection of breakpoints and abnormalities that can be related with the deregulation of critical cancer genes. The uncovering of specific patterns and the precise identification of chromosomal alteration profiles allow the research for crucial genes (oncogenes, tumor suppressor genes, and DNA repair genes) involved in the multistep process of carcinogenesis.

Due to the complexity and molecular heterogeneity of cancer genomes, the advances in cancer pathobiology have its origin on the availability of different types of experimental model systems that review the various forms of this disease. Cytogenetic approaches in rodent tumors provide valuable insights regarding the tumorigenic process and new biomarkers for clinical diagnosis and prognostic assessment, as most of the important pathways implicated in cancer are driven by orthologous genetic events present in both rodent and human tumors. The most used technologies assisting the analysis of chromosomes are the ones based in Fluorescent in situ hybridization (FISH). Each type of FISH probe has a specific ability and provides very specific answers about very specific chromosomal abnormalities.

The choice of the right probe, a pool of probes (in a multicolor approach), or a combination of FISH techniques, as well as the election of the most suitable target (from nuclei and chromosomes to extended chromatin fibers or array platforms), should be carefully evaluated as it depends on the purpose in question, the resolution needed, and the detail of the information required.

Acknowledgments The authors would like to thank Kosyakova et al. [119] and BioMed Central for reproduction of Fig. 3.4; Sandra Louzada, Susana Meles, and João Castro, all from our group, for the DNA and RNA FISH figures; and Science and Technology Foundation (FCT) from Portugal for sponsoring the PhD grant SFRH/BD/80808/2011.

References

1. Sawan C, Vaissière T, Murr R, Herceg Z. Epigenetic drivers and genetic passengers on the road to cancer. Mutat Res. 2008;642(1-2):1–13.
2. Hanahan D, Weinberg RA. Hallmarks of cancer: the next generation. Cell. 2011;144(5): 646–74.
3. Negrini S, Gorgoulis VG, Halazonetis TD. Genomic instability—an evolving hallmark of cancer. Nat Rev Mol Cell Biol. 2010;11(3):220–8.
4. McGranahan N, Burrell RA, Endesfelder D, Novelli MR, Swanton C. Cancer chromosomal instability: therapeutic and diagnostic challenges. EMBO Rep. 2012;13(6):528–38.
5. Forment JV, Kaidi A, Jackson SP. Chromothripsis and cancer: causes and consequences of chromosome shattering. Nat Rev Cancer. 2012;12(10):663–70.
6. Berger R, Bernard OA. Jumping translocations. Genes Chromosomes Cancer. 2007;46(8): 717–23.
7. Ricke RM, van Ree JH, van Deursen JM. Whole chromosome instability and cancer: a complex relationship. Trends Genet. 2008;24(9):457–66.
8. Albertson DG, Collins C, McCormick F, Gray JW. Chromosome aberrations in solid tumors. Nat Genet. 2003;34(4):369–76.
9. Micale MA. Classical and molecular cytogenetic analysis of hematolymphoid disorders. In: Eds D. Crisan, Hematopathology: genomic mechanisms of neoplastic diseases. Humana: New York; 2010. p. 39–78.
10. Stephens PJ, Greenman CD, Fu B, Yang F, Bignell GR, Mudie LJ, et al. Massive genomic rearrangement acquired in a single catastrophic event during cancer development. Cell. 2011;144(1):27–40.
11. Meyerson M, Gabriel S, Getz G. Advances in understanding cancer genomes through second-generation sequencing. Nat Rev Genet. 2010;11(10):685–96.

12. Dorritie K, Montagna C, Difilippantonio MJ, Ried T. Advanced molecular cytogenetics in human and mouse. Expert Rev Mol Diagn. 2004;4(5):663–76.
13. Thompson SL, Compton DA. Chromosomes and cancer cells. Chromosome Res. 2010;19(3):433–44.
14. Bernheim A. Cytogenomics of cancers: from chromosome to sequence. Mol Oncol. 2010;4(4):309–22.
15. Mitelman F, Johansson B, Mertens F. The impact of translocations and gene fusions on cancer causation. Nat Rev Cancer. 2007;7(4):233–45.
16. Anisimov VN, Ukraintseva SV, Yashin AI. Cancer in rodents: does it tell us about cancer in humans? Nat Rev Cancer. 2005;5(10):807–19.
17. Talmadge JE, Singh RK, Fidler IJ, Raz A. Murine models to evaluate novel and conventional therapeutic strategies for cancer. Am J Pathol. 2007;170(3):793–804.
18. Cheon D, Orsulic S. Mouse models of cancer. Annu Rev Pathol Mech Dis. 2011; 6(1):95–119.
19. Szpirer C. Cancer research in rat models. Methods Mol Biol. 2010;597:445–58.
20. International Human Genome Sequencing Consortium. Initial sequencing and analysis of the human genome. Nature. 2001;409:860–921.
21. Mouse Genome Sequencing Consortium. Initial sequencing and comparative analysis of the mouse genome. Nature. 2002;420(6915):520–62.
22. Rat Genome Sequencing Project Consortium. Genome sequence of the Brown Norway rat yields insights into mammalian evolution. Nature. 2004;428:493–521.
23. Louzada S, Adega F, Chaves R. Defining the sister rat mammary tumor cell lines HH-16 cl.2/1 and HH-16.cl.4 as an in vitro cell model for Erbb2. PLoS One. 2012;7(1), e29923.
24. Rangarajan A, Weinberg RA. Opinion: comparative biology of mouse versus human cells: modelling human cancer in mice. Nat Rev Cancer. 2003;3(12):952–9.
25. Maser RS, Choudhury B, Campbell PJ, Feng B, Wong K, Protopopov A, et al. Chromosomally unstable mouse tumours have genomic alterations similar to diverse human cancers. Nature. 2007;447(7147):966–71.
26. Huang G, Tong C, Kumbhani DS, Ashton C, Yan H, Ying Q. Beyond knockout rats: new insights into finer genome manipulation in rats. Cell Cycle. 2011;10(7):1059–66.
27. Szpirer C, Szpirer J. Mammary cancer susceptibility: human genes and rodent models. Mamm Genome. 2007;18(12):817–31.
28. Guffei A, Lichtensztejn Z, Gonçalves Dos Santos Silva A, Louis SF, Caporali A, Mai S. c-Myc-dependent formation of Robertsonian translocation chromosomes in mouse cells. Neoplasia. 2007;9(7):578–88.
29. Montagna C, Lyu M, Hunter K, Lukes L, Lowther W, Reppert T, et al. The Septin 9 (MSF) gene is amplified and overexpressed in mouse mammary gland adenocarcinomas and human breast cancer cell lines. Cancer Res. 2003;63(9):2179–87.
30. Montagna C, Andrechek ER, Padilla-Nash H, Muller WJ, Ried T. Centrosome abnormalities, recurring deletions of chromosome 4, and genomic amplification of HER2/neu define mouse mammary gland adenocarcinomas induced by mutant HER2/neu. Oncogene. 2002; 21(6):890–8.
31. Heiliger K, Hess J, Vitagliano D, Salerno P, Braselmann H, Salvatore G, et al. Novel candidate genes of thyroid tumourigenesis identified in Trk-T1 transgenic mice. Endocr Relat Cancer. 2012;19(3):409–21.
32. Deng Z, Wang Z, Xiang C, Molczan A, Baubet V, Conejo-Garcia J, et al. Formation of telomeric repeat-containing RNA (TERRA) foci in highly proliferating mouse cerebellar neuronal progenitors and medulloblastoma. J Cell Sci. 2012;125(Pt 18):4383–94.
33. Direcks WGE, van Gelder M, Lammertsma AA, Molthoff CFM. A new rat model of human breast cancer for evaluating efficacy of new anti-cancer agents in vivo. Cancer Biol Ther. 2008;7(4):532–7.
34. Sargent LM, Senft JR, Lowry DT, Jefferson AM, Tyson FL, Malkinson AM, et al. Specific chromosomal aberrations in mouse lung adenocarcinoma cell lines detected by spectral

karyotyping: a comparison with human lung adenocarcinoma. Cancer Res. 2002; 62(4):1152–7.

35. Meuwissen R, Berns A. Mouse models for human lung cancer. Genes Dev. 2005; 19(6):643–64.

36. Logan JA. Chromosomes in the mouse. Science. 1908;27:443–4.

37. Painter TS. The chromosomes of rodents. Science. 1926;64:336.

38. Tjio JH, Levan A. The chromosome number in man. Hereditas. 1956;42:1–6.

39. Vrba M, Donner L. Chromosome numbers and karyotypes of two rat tumours induced by rous sarcoma virus in vitro. Folia Biol. 1964;10:373–80.

40. Winge Ö. Zytologische Untersuchungen über die Natur maligner Tumoren. Z Zelforsch Mikrosk Anat. 1930;10(4):683–735.

41. Hsu TC. Human and mammalian cytogenetics: an historical perspective. New York: Springer; 1979. 186pp.

42. Caspersson T, Zech L, Johansson C. Differential binding of alkylating fluorochromes in human chromosomes. Exp Cell Res. 1970;60(3):315–9.

43. Shepard JS, Wurster-Hill DH, Pettengill OS, Sorenson GD. Giemsa-banded chromosomes of mouse myeloma in relationship to oncogenicity. Cytogenet Cell Genet. 1974;13(3): 279–309.

44. Crews S, Barth R, Hood L, Prehn J, Calame K. Mouse c-myc oncogene is located on chromosome 15 and translocated to chromosome 12 in plasmacytomas. Science. 1982;218(4579): 1319–21.

45. Adhvaryu S. Chromosome analysis in cancer patients: applications and limitations. In: Eds Günter Obe, Vijayalaxmi, Chromosomal alterations. Berlin: Springer; 2007. p. 479–93.

46. Ried T, Difilippantonio M. Characterization of chromosomal translocations in mouse models of hematological malignancies using spectral karyotyping, FISH, and immunocytochemistry. In: Eds Jeffrey E. Green, Thomas Ried, Genetically engineered mice for cancer research. New York: Springer; 2012. p. 193–207.

47. Speicher MR, Carter NP. The new cytogenetics: blurring the boundaries with molecular biology. Nat Rev Genet. 2005;6(10):782–92.

48. Fabris VT, Lodillinsky C, Pampena MB, Belgorosky D, Lanari C, Eiján AM. Cytogenetic characterization of the murine bladder cancer model MB49 and the derived invasive line MB49-I. Cancer Genet. 2012;205(4):168–76.

49. Pinkel D, Straume T, Gray JW. Cytogenetic analysis using quantitative, high-sensitivity, fluorescence hybridization. Proc Natl Acad Sci U S A. 1986;83:2934–8.

50. Cremer T, Landegent J, Bruckner A, Scholl HP, Schardin M, Hager HD, et al. Detection of chromosome aberrations in the human inter- phase nucleus by visualization of specific target DNA's with radioactive and non-radioactive in situ hybridization techniques: diagnosis of trisomy 18 with probe L1.84. Hum Genet. 1986;74:346–52.

51. Swanton C. Intratumor heterogeneity: evolution through space and time. Cancer Res. 2012;72(19):4875–82.

52. Gerlinger M, Rowan AJ, Horswell S, Larkin J, Endesfelder D, Gronroos E, et al. Intratumor heterogeneity and branched evolution revealed by multiregion sequencing. N Engl J Med. 2012;366(10):883–92.

53. Weier HG, Greulich-Bode KM, Ito Y, Lersch RA, Fung J. FISH in cancer diagnosis and prognostication: from cause to course of disease. Expert Rev Mol Diagn. 2002;2(2):109–19.

54. Tönnies H. Modern molecular cytogenetic techniques in genetic diagnostics. Trends Mol Med. 2002;8(6):246–50.

55. Ried T, Liyanage M, du Manoir S, Heselmeyer K, Auer G, Macville M, et al. Tumor cytogenetics revisited: comparative genomic hybridization and spectral karyotyping. J f Mol Med. 1997;75(11–12):801–14.

56. Miura Y, Keira Y, Ogino J, Nakanishi K, Noguchi H, Inoue T, et al. Detection of specific genetic abnormalities by fluorescence in situ hybridization in soft tissue tumors. Pathol Int. 2011;62(1):16–27.

57. Brandal P, Micci F, Bjerkehagen B, Eknæs M, Larramendy M, Lothe RA, et al. Molecular cytogenetic characterization of desmoid tumors. Cancer Genet Cytogenet. 2003;146(1):1–7.
58. Heng H. Released chromatin or DNA fiber preparations for high-resolution fiber FISH. In: Eds I.A. Darby, *In situ* hybridization protocols, vol. 123. Humana: New Jersey: Humana Press; 2000. p. 69–81.
59. Wang N. Methodologies in cancer cytogenetics and molecular cytogenetics. Am J Med Genet. 2002;115(3):118–24.
60. Pinkel D, Segraves R, Sudar D, Clark S, Poole I, Kowbel D, et al. High resolution analysis of DNA copy number variation using comparative genomic hybridization to microarrays. Nat Genet. 1998;20(2):207–11.
61. Chung Y, Jonkers J, Kitson H, Fiegler H, Humphray S, Scott C, et al. A whole-genome mouse BAC microarray with 1-Mb resolution for analysis of DNA copy number changes by array comparative genomic hybridization. Genome Res. 2004;14(1):188–96.
62. Shaffer LG. Microarray-based cytogenetics. In: Eds Steven L. Gersen, Martha B. Keagle, The principles of clinical cytogenetics. New York: Springer; 2013. p. 441–50.
63. Cho EK. Array-based comparative genomic hybridization and its application to cancer genomes and human genetics. J Lung Cancer. 2011;10(2):77–86.
64. Silva AG, Graves HA, Guffei A, Ricca TI, Mortara RA, Jasiulionis MG, Mai S. Telomere-centromere driven genomic instability contributes to karyotype evolution in a mouse model of melanoma. Neoplasia. 2010;12(1):11–9.
65. Liu X, Sempere LF, Ouyang H, Memoli VA, Andrew AS, Luo Y, et al. MicroRNA-31 functions as an oncogenic microRNA in mouse and human lung cancer cells by repressing specific tumor suppressors. J Clin Invest. 2010;120(4):1298–309.
66. Lu J, Tsourkas A. Imaging individual microRNAs in single mammalian cells in situ. Nucleic Acids Res. 2009;37(14), e100.
67. Adamovic T, Trossö F, Roshani L, Andersson L, Petersen G, Rajaei S, et al. Oncogene amplification in the proximal part of chromosome 6 in rat endometrial adenocarcinoma as revealed by combined BAC/PAC FISH, chromosome painting, zoo-FISH, and allelotyping. Genes Chromosomes Cancer. 2005,44(2).139–53.
68. Karlsson A, Helou K, Walentinsson A, Hedrich HJ, Szpirer C, Levan G. Amplification of Mycn, Ddx1, Rrm2, and Odc1 in rat uterine endometrial carcinomas. Genes Chromosomes Cancer. 2001;31(4):345–56.
69. Ried T, Schröck E, Ning Y, Wienberg J. Chromosome painting: a useful art. Hum Mol Genet. 1998;7(10):1619–26.
70. Telenius H, Carter NP, Bebb CE, Nordenskjöld M, Ponder BA, Tunnacliffe A. Degenerate oligonucleotide-primed PCR: general amplification of target DNA by a single degenerate primer. Genomics. 1992;13(3):718–25.
71. Guan XY, Zhang H, Bittner M, Jiang Y, Meltzer P, Trent J. Chromosome arm painting probes. Nat Genet. 1996;12(1):10–1.
72. Guan XY, Trent JM, Meltzer PS. Generation of band-specific painting probes from a single microdissected chromosome. Hum Mol Genet. 1993;2(8):1117–21.
73. Doležel J, Vrána J, Šafář J, Bartoš J, Kubaláková M, Simková H. Chromosomes in the flow to simplify genome analysis. Funct Integr Genomics. 2012;12(3):397–416.
74. Adega F, Chaves R, Kofler A, Krausman PR, Masabanda J, Wienberg J, et al. High-resolution comparative chromosome painting in the Arizona collared peccary (Pecari tajacu, Tayassuidae): a comparison with the karyotype of pig and sheep. Chromosome Res. 2006;14(3):243–51.
75. Liechty MC, Hall BK, Scalzi JM, Davis LM, Caspary WJ, Hozier JC. Mouse chromosome-specific painting probes generated from microdissected chromosomes. Mamm Genome. 1995;6(9):592–4.
76. Yang F, Trifonov V, Ng BL, Kosyakova N, Carter NP. Generation of paint probes by flow-sorted and microdissected chromosomes. In: Eds Thomas Liehr, Fluorescence in situ hybridization (FISH)—application guide. Berlin: Springer; 2009. p. 35–52.

77. Louzada S, Paço A, Kubickova S, Adega F, Guedes-Pinto H, Rubes J, et al. Different evolutionary trails in the related genomes Cricetus cricetus and Peromyscus eremicus (Rodentia, Cricetidae) uncovered by orthologous satellite DNA repositioning. Micron. 2008;39(8):1149–55.

78. Meltzer PS, Guan XY, Burgess A, Trent JM. Rapid generation of region specific probes by chromosome microdissection and their application. Nat Genet. 1992;1(1):24–8.

79. Carter NP, Ferguson-Smith MA, Perryman MT, Telenius H, Pelmear AH, Leversha MA, et al. Reverse chromosome painting: a method for the rapid analysis of aberrant chromosomes in clinical cytogenetics. J Med Genet. 1992;29(5):299–307.

80. Coleman AE, Kovalchuk AL, Janz S, Palini A, Ried T. Jumping translocation breakpoint regions lead to amplification of rearranged Myc. Blood. 1999;93(12):4442–4.

81. Speicher MR, Gwyn Ballard S, Ward DC. Karyotyping human chromosomes by combinatorial multi-fluor FISH. Nat Genet. 1996;12(4):368–75.

82. Liyanage M, Coleman A, du Manoir S, Veldman T, McCormack S, Dickson RB, et al. Multicolour spectral karyotyping of mouse chromosomes. Nat Genet. 1996;14(3):312–5.

83. Tanke HJ, Wiegant J, van Gijlswijk RP, Bezrookove V, Pattenier H, Heetebrij RJ, et al. New strategy for multi-colour fluorescence in situ hybridisation: COBRA: COmbined Binary RAtio labelling. Eur J Hum Genet. 1999;7(1):2–11.

84. Bayani JM, Squire JA. Applications of SKY in cancer cytogenetics. Cancer Invest. 2002;20(3):373–86.

85. Macville M, Veldman T, Padilla-Nash H, Wangsa D, O'Brien P, Schröck E, et al. Spectral karyotyping, a 24-colour FISH technique for the identification of chromosomal rearrangements. Histochem Cell Biol. 1997;108(4-5):299–305.

86. Falck E, Hedberg C, Klinga-Levan K, Behboudi A. SKY analysis revealed recurrent numerical and structural chromosome changes in BDII rat endometrial carcinomas. Cancer Cell Int. 2011;11(1):20.

87. Buwe A, Steinlein C, Koehler MR, Bar-Am I, Katzin N, Schmid M. Multicolor spectral karyotyping of rat chromosomes. Cytogenet Genome Res. 2003;103(1-2):163–8.

88. Vajen B, Modlich U, Schienke A, Wolf S, Skawran B, Hofmann W, et al. Histone methyltransferase Suv39h1 deficiency prevents Myc-induced chromosomal instability in murine myeloid leukemias. Genes Chromosomes Cancer. 2013;52(4):423–30.

89. Weaver Z, Montagna C, Xu X, Howard T, Gadina M, Brodie SG, et al. Mammary tumors in mice conditionally mutant for Brca1 exhibit gross genomic instability and centrosome amplification yet display a recurring distribution of genomic imbalances that is similar to human breast cancer. Oncogene. 2002;21(33):5097–107.

90. Coleman A, Schrock E, Weaver Z, du Manoir S, Yang F, Ferguson-Smith M, et al. Previously hidden chromosome aberrations in T (12; 15)-positive BALB/c plasmacytomas uncovered by multicolor spectral karyotyping. Cancer Res. 1997;57:4585–92.

91. Karst C, Trifonov V, Romanenko SA, Claussen U, Mrasek K, Michel S, et al. Molecular cytogenetic characterization of the mouse cell line WMP2 by spectral karyotyping and multicolor banding applying murine probes. Int J Mol Med. 2006;17(2):209–13.

92. Kallioniemi A, Kallioniemi OP, Sudar D, Rutovitz D, Gray JW, Waldman F, et al. Comparative genomic hybridization for molecular cytogenetic analysis of solid tumors. Science. 1992;258(5083):818–21.

93. Christian AT, Snyderwine EG, Tucker JD. Comparative genomic hybridization analysis of PhIP-induced mammary carcinomas in rats reveals a cytogenetic signature. Mutat Res. 2002;506–507:113–9.

94. Pinkel D, Albertson DG. Comparative genomic hybridization. Annu Rev Genomics Hum Genet. 2005;6(1):331–54.

95. Salawu A, Ul-Hassan A, Hammond D, Fernando M, Reed M, Sisley K. High quality genomic copy number data from archival formalin-fixed paraffin-embedded leiomyosarcoma: optimisation of universal linkage system labelling. PLoS One. 2012;7(11), e50415.

96. Schmidt-Kittler O, Ragg T, Daskalakis A, Granzow M, Ahr A, Blankenstein TJF, et al. From latent disseminated cells to overt metastasis: genetic analysis of systemic breast cancer progression. Proc Natl Acad Sci U S A. 2003;100(13):7737–42.

97. Valli R, Marletta C, Pressato B, Montalbano G, lo Curto F, Pasquali F, et al. Comparative genomic hybridization on microarray (a-CGH) in constitutional and acquired mosaicism may detect as low as 8% abnormal cells. Mol Cytogenet. 2011;4(1):13.

98. Kogan SC, Brown DE, Shultz DB, Truong BT, Lallemand-Breitenbach V, Guillemin MC, et al. BCL-2 cooperates with promyelocytic leukemia retinoic acid receptor alpha chimeric protein (PMLRARalpha) to block neutrophil differentiation and initiate acute leukemia. J Exp Med. 2001;193(4):531–43.

99. Kallioniemi A, Kallioniemi OP, Piper J, Tanner M, Stokke T, Chen L, et al. Detection and mapping of amplified DNA sequences in breast cancer by comparative genomic hybridization. Proc Natl Acad Sci U S A. 1994;91(6):2156–60.

100. Zheng W, Gustafson DR, Sinha R, Cerhan JR, Moore D, Hong CP, et al. Well-done meat intake and the risk of breast cancer. J Natl Cancer Inst. 1998;90(22):1724–9.

101. Climent J, Garcia JL, Mao JH, Arsuaga J, Perez-Losada J. Characterization of breast cancer by array comparative genomic hybridization. Biochem Cell Biol. 2007;85(4):497–508. This paper is one of a selection of papers published in this Special Issue, entitled 28th International West Coast Chromatin and Chromosome Conference, and has undergone the Journal's usual peer review process.

102. Solinas-Toldo S, Lampel S, Stilgenbauer S, Nickolenko J, Benner A, Döhner H, et al. Matrix-based comparative genomic hybridization: biochips to screen for genomic imbalances. Genes Chromosomes Cancer. 1997;20(4):399–407.

103. van den Ijssel P, Tijssen M, Chin S, Eijk P, Carvalho B, Hopmans E, et al. Human and mouse oligonucleotide-based array CGH. Nucleic Acids Res. 2005;33(22), e192.

104. Zhao X, Li C, Paez JG, Chin K, Jänne PA, Chen T, et al. An integrated view of copy number and allelic alterations in the cancer genome using single nucleotide polymorphism arrays. Cancer Res. 2004;64(9):3060–71.

105. Hodgson G, Hager JH, Volik S, Hariono S, Wernick M, Moore D, et al. Genome scanning with array CGH delineates regional alterations in mouse islet carcinomas. Nat Genet. 2001;29(4):459–64.

106. Hackett CS, Hodgson JG, Law ME, Fridlyand J, Osoegawa K, de Jong PJ, et al. Genome-wide array CGH analysis of murine neuroblastoma reveals distinct genomic aberrations which parallel those in human tumors. Cancer Res. 2003;63(17):5266–73.

107. Sander S, Bullinger L, Karlsson A, Giuriato S, Hernandez-Boussard T, Felsher DW, et al. Comparative genomic hybridization on mouse cDNA microarrays and its application to a murine lymphoma model. Oncogene. 2005;24(40):6101–7.

108. van de Wiel MA, Costa JL, Smid K, Oudejans CBM, Bergman AM, Meijer GA, et al. Expression microarray analysis and oligo array comparative genomic hybridization of acquired gemcitabine resistance in mouse colon reveals selection for chromosomal aberrations. Cancer Res. 2005;65(22):10208–13.

109. Yi Y, Nandana S, Case T, Nelson C, Radmilovic T, Matusik RJ, et al. Candidate metastasis suppressor genes uncovered by array comparative genomic hybridization in a mouse allograft model of prostate cancer. Mol Cytogenet. 2009;2(1):18.

110. To MD, Quigley DA, Mao JH, del Rosario R, Hsu J, Hodgson G, et al. Progressive genomic instability in the FVB/KrasLA2 mouse model of lung cancer. Mol Cancer Res. 2011;9(10):1339–45.

111. Kazmi SJ, Byer SJ, Eckert JM, Turk AN, Huijbregts RPH, Brossier NM, et al. Transgenic mice overexpressing neuregulin-1 model neurofibroma-malignant peripheral nerve sheath tumor progression and implicate specific chromosomal copy number variations in tumorigenesis. Am J Pathol. 2013;182(3):646–67.

112. Samuelson E, Karlsson S, Partheen K, Nilsson S, Szpirer C, Behboudi A. BAC CGH-array identified specific small-scale genomic imbalances in diploid DMBA-induced rat mammary tumors. BMC Cancer. 2012;12:352.

113. Fiegler H, Gribble SM, Burford DC, Carr P, Prigmore E, Porter KM, et al. Array painting: a method for the rapid analysis of aberrant chromosomes using DNA microarrays. J Med Genet. 2003;40(9):664–70.

114. Gribble SM, Ng BL, Prigmore E, Fitzgerald T, Carter NP. Array painting: a protocol for the rapid analysis of aberrant chromosomes using DNA microarrays. Nat Protoc. 2009; 4(12):1722–36.

115. Baptista J, Mercer C, Prigmore E, Gribble SM, Carter NP, Maloney V, et al. Breakpoint mapping and array CGH in translocations: comparison of a phenotypically normal and an abnormal cohort. Am J Hum Genet. 2008;82(4):927–36.

116. Chudoba I, Plesch A, Lörch T, Lemke J, Claussen U, Senger G. High resolution multicolorbanding: a new technique for refined FISH analysis of human chromosomes. Cytogenet Cell Genet. 1999;84(3-4):156–60.

117. Liehr T, Gross M, Karst C, Glaser M, Mrasek K, Starke H, et al. FISH banding in tumor cytogenetics. Cancer Genet Cytogenet. 2006;164(1):88–9.

118. Liehr T, Weise A, Hinreiner S, Mkrtchyan H, Mrasek K, Kosyakova N. Characterization of chromosomal rearrangements using multicolor-banding (MCB/m-band). In: Eds J.M. Bridger, K. Morris, Fluorescence in situ hybridization (FISH). Humana: New York: Humana Press; 2010. p. 231–8.

119. Kosyakova N, Hamid AB, Chaveerach A, Pinthong K, Siripiyasing P, Supiwong W, et al. Generation of multicolor banding probes for chromosomes of different species. Mol Cytogenet. 2013;6(1):1–1.

120. Trifonov VA, Kosyakova N, Romanenko SA, Stanyon R, Graphodatsky AS, Liehr T. New insights into the karyotypic evolution in muroid rodents revealed by multicolor banding applying murine probes. Chromosome Res. 2010;18(2):265–75.

121. Leibiger C, Kosyakova N, Mkrtchyan H, Glei M, Trifonov V, Liehr T. First molecular cytogenetic high resolution characterization of the NIH 3T3 cell line by murine multicolor banding. J Histochem Cytochem. 2013;61(4):306–12.

122. Benedeck K, Chudoba I, Klein G, Wiener F, Mai S. Rearrangements of the telomeric region of mouse chromosome 11 in Pre-B ABL/MYC cells revealed by mBANDing, spectral karyotyping, and fluorescence in-situ hybridization with a subtelomeric probe. Chromosome Res. 2004;12:777–85.

123. Liechty MC, Carpio CM, Aytay S, Clase AC, Puschus KL, Sims KR, et al. Hybridizationbased karyotyping of mouse chromosomes: hybridization-bands. Cytogenet Cell Genet. 1999;86(1):34–8.

124. Henegariu O, Dunai J, Chen XN, Korenberg JR, Ward DC, Greally JM. A triple color FISH technique for mouse chromosome identification. Mamm Genome. 2001;12(6):462–5.

125. Müller S, O'Brien PC, Ferguson-Smith MA, Wienberg J. Cross-species colour segmenting: a novel tool in human karyotype analysis. Cytometry. 1998;33(4):445–52.

126. Müller S, Wienberg J. Multicolor chromosome bar codes. Cytogenet Genome Res. 2006;114(3-4):245–9.

127. Gonçalves Dos Santos Silva A, Sarkar R, Harizanova J, Guffei A, Mowat M, Garini Y, et al. Centromeres in cell division, evolution, nuclear organization and disease. J Cell Biochem. 2008;104(6):2040–58.

128. Murnane JP. Telomere dysfunction and chromosome instability. Mutat Res. 2012; 730(1-2):28–36.

129. Hoebee B, de Stoppelaar JM, Suijkerbuijk RF, Monard S. Isolation of rat chromosomespecific paint probes by bivariate flow sorting followed by degenerate oligonucleotide primed-PCR. Cytogenet Cell Genet. 1994;66(4):277–82.

130. Doherty AT, Ellard S, Parry EM, Parry JM. A study of the aneugenic activity of trichlorfon detected by centromere-specific probes in human lymphoblastoid cell lines. Mutat Res. 1996;372(2):221–31.

131. Blasco MA, Lee HW, Hande MP, Samper E, Lansdorp PM, DePinho RA, et al. Telomere shortening and tumor formation by mouse cells lacking telomerase RNA. Cell. 1997;91(1):25–34.

132. Rudolph KL, Millard M, Bosenberg MW, DePinho RA. Telomere dysfunction and evolution of intestinal carcinoma in mice and humans. Nat Genet. 2001;28(2):155–9.

133. Begus-Nahrmann Y, Hartmann D, Kraus J, Eshraghi P, Scheffold A, Grieb M, et al. Transient telomere dysfunction induces chromosomal instability and promotes carcinogenesis. J Clin Invest. 2012;122(6):2283–8.
134. Sarkar R, Guffei A, Vermolen BJ, Garini Y, Mai S. Alterations of centromere positions in nuclei of immortalized and malignant mouse lymphocytes. Cytometry A. 2007;71(6):386–92.
135. Le Beau MM, Bitts S, Davis EM, Kogan SC. Recurring chromosomal abnormalities in leukemia in PML-RARA transgenic mice parallel human acute promyelocytic leukemia. Blood. 2002;99(8):2985–91.
136. Le Beau MM, Davis EM, Patel B, Phan VT, Sohal J, Kogan SC. Recurring chromosomal abnormalities in leukemia in PML-RARA transgenic mice identify cooperating events and genetic pathways to acute promyelocytic leukemia. Blood. 2003;102(3):1072–4.
137. Machluf Y, Levkowitz G. Visualization of mRNA expression in the zebrafish embryo. In: Eds J.E. Gerst, RNA detection and visualization, vol. 714. Humana: New York: Humana Press; 2011. p. 83–102.
138. Lécuyer E. High resolution fluorescent in situ hybridization in Drosophila. In: Eds J.E. Gerst, RNA Detection and visualization, vol. 714. Humana: New York: Humana Press; 2011. p. 31–47.
139. Tam R, Shopland LS, Johnson von C, McNeil JA, Lawrence AJB. Applications of RNA FISH for visualizing gene expression and nuclear architecture. In: Eds Barbara Beatty, Sabine Mai, Jeremy Squire, FISH: New York: Oxford University Press; 2002. p. 93.
140. Itzkovitz S, van Oudenaarden A. Validating transcripts with probes and imaging technology. Nat Methods. 2011;8(4s):S12–9.
141. Ting DT, Lipson D, Paul S, Brannigan BW, Akhavanfard S, Coffman EJ, et al. Aberrant overexpression of satellite repeats in pancreatic and other epithelial cancers. Science. 2011;331(6017):593–6.
142. Yildirim E, Kirby JE, Brown DE, Mercier FE, Sadreyev RI, Scadden DT, et al. Xist RNA is a potent suppressor of hematologic cancer in mice. Cell. 2013;152(4):727–42.
143. Ganesan S, Silver DP, Greenberg RA, Avni D, Drapkin R, Miron A, et al. BRCA1 supports XIST RNA concentration on the inactive X chromosome. Cell. 2002;111(3):393–405.
144. Azzalin CM, Reichenbach P, Khoriauli L, Giulotto E, Lingner J. Telomeric repeat containing RNA and RNA surveillance factors at mammalian chromosome ends. Science. 2007;318(5851):798–801.

Chapter 4
Quantitation of Immunohistochemistry by Image Analysis Technique

Klaus Kayser and Gian Kayser

Introduction

In 1941 Coons reported the detection of antigens in deep frozen human tissue cut on glass slides and incubated with "antibody solution" [1]. The reported technique to detect antigens in thin tissue sections was handicapped by the inappropriate technique to store and analyse fluorescent signals at that time. The breakthrough and practical application of immunohistochemistry (IHC) started in 1974/1975, when Taylor and Burns as well as Kohler and Milstein published their technology of visualisation of antigen–antibody reaction by diaminobenzidine (DAB) horseradish peroxidase reaction, and the development of hybridoma cell lines in order to obtain suitable antibodies [2–7]. Since that time IHC has been matured to a mandatory tool in routine diagnostic surgical pathology, and consecutively, several diseases are classified according to the expression of specific antigens such as the CD (core density) family in malignant lymphomas or that of specific cytoskeletons in carcinomas and sarcomas [8–10].

The principle of IHC is shown in Fig. 4.1. It is based upon specific binding capacities of macromolecules (antibodies) to other macromolecules (antigens). After purification of antigens, the specific antibodies can be obtained either after injection of the antigens into suitable animal models (mice, creation of so-called polyclonal antibodies which possess several epitopes) or by the creation of specific hybridoma cell lines that continuously release the wanted (monoclonal) antibody, i.e. a macromolecule that possesses only one specific epitope to the antigen.

K. Kayser (✉)
Institute of Pathology, Charite, Berlin, Berlin, Germany
e-mail: Klaus.kayser@charite.de

G. Kayser
Institute of Pathology, University of Freiburg, Freiburg, Germany
e-mail: gian.kayser@uniklinik-freiburg.de

© Springer Science+Business Media New York 2016   51
S.A. Aziz, R. Mehta (eds.), *Technical Aspects of Toxicological Immunohistochemistry*, DOI 10.1007/978-1-4939-1516-3_4

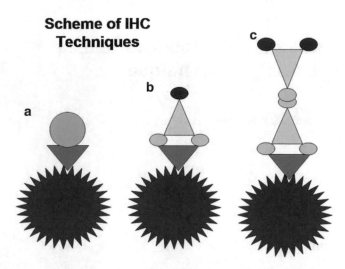

Fig. 4.1 Scheme of IHC: (**a**) direct visualisation (usually with fluorescent dye); (**b**) normal AP technique; (**c**) APAAP and ABC technique with multiple visualisation dyes

The bound antibodies can be visualised by two different techniques: (a) by a secondary immune complex that recognises the (biotin) labelled antibody or (b) by a fluorescent signal that is released either directly (bound to the antibody) or indirectly (by a secondary fluorescent labelled antibody directed to a so-called linker molecule (biotin)) [11–14]. An enhancement of the indirect (two step method) has been developed in adding a third layer of antibodies to the second layer (first antibody) that are labelled with two (or more) visualisation molecules (so-called APAAP technique). Today, most laboratories use the avidin/streptavidin-biotin technique, because streptavidin and avidin possess strong binding capacities for biotin (ABC technique). These molecules can be labelled with several enzyme reaction molecules. The sensitivity of the antigen–antibody reaction is enhanced for about 6–10 times using this technology [10].

It is important to distinguish between the two visualisation techniques as they follow different natural laws of light propagation: Conventional transmission light microscopy uses the absorption of light induced by the applied stains. Conventional IHC visualises the antibodies in brown (DAB) or red (AP) or black (silver) colours. The intensity of these stains decreases logarithmically in relation to the molecules' (antibody) density which is automatically corrected by the human eye [15]. Therefore, human-based light absorbance evaluation disregards the logarithmic law. This feature has to be taken into account when quantitative measurements are undertaken [16]. On the other hand, fluorescent signals are linear related to the signal density and do not require a logarithmic "correction".

Tissue Microarrays

The significant frequency of IHC application in addition to the numerous available antibodies has forced the intensive search for more efficient histological techniques as just to apply one antibody to one glass slide. The result is the development of arrays which are pre-prepared matrices of small tissue punches that can be analysed with a broad number of different antibodies at the same time [17, 18]. An example of such an arrangement is shown in Fig. 4.2. An exemplarily tissue cut is analysed microscopically, and the corresponding areas of the associated tissue block are punched out and embedded in a new tissue block in a matrix-like arrangement. Such tissue microarrays (TMA) can contain up to several hundred punches, which might be obtained either from the same tumour (and then be exposed to a set of different probes) or from a series of different tumours (and then be exposed to the same probe) [19–22].

The probes can differ, and include antibodies applicable in IHC, DNA, or RNA segments applicable in hybridisation or PCR, or other molecular genetic methods.

The analysis of TMA requires an automated or at least semi-automated measurement of the signals, followed by detailed statistics. An open question remains as the issue of disease heterogeneity combined with the small size of the analysed probes has not been solved to our knowledge [18–20, 22–24]. The smaller the size of the examined tissue, the more it is subject for individual alterations within the lesion which might be of significance for the individual patient; however, it might not reflect to a more generalised classification, development, or therapeutic response.

Example of Tissue Micro Array (TMA)

Breast carcinoma cases of 18 * 8 (144) spots

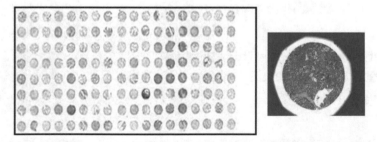

Glass slide of TMA spots   Individual spot

Fig. 4.2 Example of a TMA with 144 tissue spots of a breast carcinoma

Tissue-Based Diagnosis and Types of Diagnoses

There is no doubt that surgical pathology is the medical diagnostic technique with the highest diagnostic accuracy in terms of sensitivity and specificity [25, 26]. The mentioned high intra- and inter-observer discrepancies are usually related to highly differentiated morphological patterns and do not possess any or only weak clinical, i.e. therapeutic impact [27–33]. They reflect to the broad variance of morphological appearance in most of the living systems, at least in histology [34–36]. The relationship of morphology to functional properties such as occurrence of malignant growth is only crudely understood in our opinion.

Despite this constraint, morphologically based diagnoses have been proven to be a solid basis in several diseases such as cancer or specific infections (tuberculosis etc.). These diagnoses are derived from conventionally stained (H&E) slides, and are called conventional diagnosis [37–39]. As shown in Fig. 4.3, at least four different "subtypes" of tissue-based diagnosis (judged from surgical pathology specimens) can be distinguished. In addition to the conventional diagnosis, specific commonly IHC founded diagnoses can be histologically derived from surgical specimens. They can be associated with the patient's diagnosis, and are called prognostic diagnoses.

Diagnosis types and computerized assistance (diagnosis assistants)

- Classic diagnosis: primary: not essential; secondary: partly

- Prognosis estimation (quantitative immunohistochemistry): partly

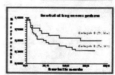

- Predictive diagnosis (quantitative immunohistochemistry, gene analysis): partly

- Risk estimation (array technique): essential

Fig. 4.3 Survey of the different diagnosis types

Her2_neu breast carcinoma / entropy

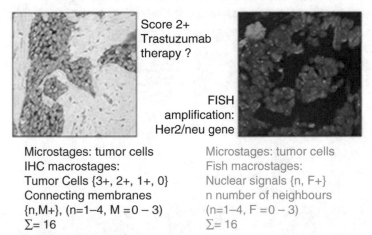

Score 2+
Trastuzumab
therapy ?

FISH
amplification:
Her2/neu gene

Microstages: tumor cells
IHC macrostages:
Tumor Cells {3+, 2+, 1+, 0}
Connecting membranes
{n,M+}, (n=1–4, M =0 – 3)
Σ= 16

Microstages: tumor cells
Fish macrostages:
Nuclear signals {n, F+}
n number of neighbours
(n=1–4, F =0 – 3)
Σ= 16

Fig. 4.4 IHC of the herceptin receptor, the corresponding FISH signals, and entropy calculations in respect of micro- and macrostages

IHC can visualise cellular and functional properties such as percentage of proliferating tumour cells, expression of factors that induce/promote vascularisation, and other properties that might obviously be of influence on tumour propagation and metastasis. An example of such a constellation is depicted in Fig. 4.4. The therapeutic ill-defined Her2-neu score of 2+ in breast carcinomas can be further refined by entropy analysis.

Whereas prognostic diagnoses are obtained from IHC analyses of extracellular and intercellular (structural) features, predictive diagnoses are based upon the investigations of associated intracellular pathways. The most frequently analysed pathways are related to the expression of the epidermal growth factor receptor (EGFR), a molecule that can promote growth-related genes (oncogenes) via specific associated genes (kras, etc.). A generally accepted model of some of these pathways is shown in Fig. 4.5 (after [40]). The blocking of the EGFR by use of specific antibodies will only work if the intracellular pathways (and their genes) are not altered, i.e. the obtained constellation is of predictive value for potential therapies.

The described IHC (and molecular genetic applications such as PCR, DNA sequencing, etc.) are all applied to detect diseases, i.e. once the cancer, infection, or inflammation became manifest. The last kind of disease diagnosis is the risk estimation diagnosis, which describes the risk of a population to develop a certain disease. An example of such a diagnosis is the expression of the BRCA1 (2) gene in women associated with the (increased) risk of breast cancer [41–43].

With the exception of the classic diagnosis, all other diagnosis types require at least some semi-quantitative (scores) if not a complete quantification of the visualised signals (see Fig. 4.3).

Molecular Marker – EML4-ALK1 fusion gene (echinoderm microtubule-associated protein-like 4 (*EML4*) gene and anaplastic lymphoma kinase (*ALK*)

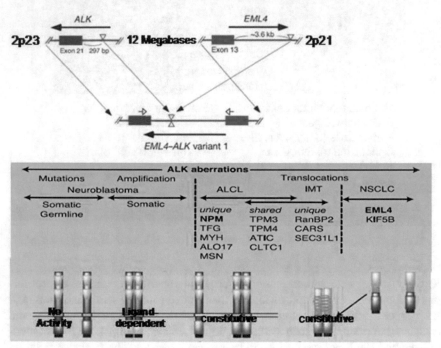

Fig. 4.5 Scheme of EML4-ALK1 fusion gene and the associated protein expression. Soda et al. Nature. 2007;448:561–7 [40]

Basics of Image Analysis

Images are one important source of medical information. These include whole body live imaging (X-rays, computerised technology (CT), nuclear resonance imaging (NMR), radio-isotopes, ultrasound, and others. Radiological images provide the clinician with information about the extension of a suggested disease in contrast to tissue-based diagnosis which is mainly dealing with its (detailed) nature (diagnosis). The general procedure on how to obtain a tissue-based diagnosis is shown in Fig. 4.6. Biopsies precede the analysis of surgical specimen, and a quality evaluation of the surgical treatment (usual analysis of resection boundaries) adds to the establishment of tissue-based diagnosis [44].

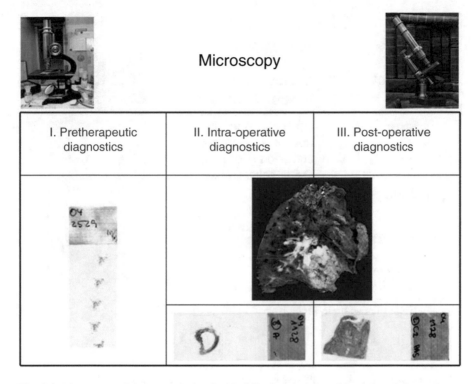

Fig. 4.6 Diagnoses and therapy-associated with different schedules of tissue based diagnosis

Principles of Microscopic Image Analysis

Each digitised image consists of a matrix of grey values (pixels) distributed on three colour dimensions (most frequently in red, green, blue (rgb)) axes. Thus, the information mapped on this three-dimensional colour space is a function of the grey values in principle. The information acquisition by any viewer (pathologist) can be formulated by a function that transfers the grey value distribution to the analysing pathologist and depends in addition upon the pathologist's knowledge and interpretation. Interpretation requires in addition the detection and identification of objects, which represent the "external" knowledge. At light microscopy, magnification objects are nuclei and cells, cellular agglutinations arranged to functional units such as vessels, glands, nerves, etc., or intestinal surfaces, or skin. The neighbourhood of these functional units or biological meaningful objects may contain a biological function again, for example goblet cells and ciliated cells in bronchial mucosa.

A more sophisticated analysis directs to the concept of ordered structures in living organisms which seem to be a necessity to stay alive [34, 35, 45–48].

Object – Structure - Texture

Feature	Biological unit	
• Object	nucleus, cell, vessel, etc.	
• Structure	clusters, dependent upon neighborhood condition (Voronoi, O'Callaghan, etc.)	
• Texture	pixel based, dependent upon applied transformation.	

> Remarks: Objects have to be predefined and can from regular arrengements, which can serve for objects at a different (lower) magnification (level of strutures).

Fig. 4.7 Explanation of the main components of image content information

In respect to image analysis, the derivative is the idea that all content-based image information can be derived from three different, not necessarily independent, parameters, namely texture, object, and structure [37, 49, 50]. The principle is demonstrated in Fig. 4.7. Texture is a mapping of pixel-based grey values, which can be transformed by different transformations such as Fourier analysis, regressive analysis comparable to time series analysis, Laplace transformations, local and other filters, etc. [36, 37, 51, 52]. The second and third techniques depend on segmentation and identification of external defined objects and their spatial relationship. The definition of a neighbourhood is mandatory to analyse the structures built by the objects. Voronoi's neighbourhood condition and the graph theory are often applied to measure structural features [15, 53–58].

The three image information components can be set in order as follows:

1. An image texture is a prerequisite for the presence of objects and structures, i.e. without an image texture no objects and consecutively no structures do exist.
2. Textures basically depend upon the image size and the chosen start point of analysis. Most algorithms require an image size $>256 \times 256$ pixels for reliable performance.
3. Objects can only be segmented if the grey value distribution of the image displays at least one maximum at a statistically significant level (to be separated from noise).
4. Objects are a prerequisite for structure(s), i.e. without any object no structure.

5. All object features can be derived from its size (in pixels) and its internal grey value distribution (including its surface). Thus, objects have to possess a minimum size in order to perform feature extraction at a statistically significant level. The average size of an object should amount to 1000 pixels, and a minimum of 100 objects should be measured.
6. The average number of objects that form a structure should amount to 100 objects.

Colour Space

Different colour spaces have been defined to present and analyse a digitised image. The most frequently used one is the image presentation in the rgb space. The rgb space takes into account that each colour can be composed by contribution of colours red, green, and blue. The rgb space is suitable to working with fluorescent images because most of the signals are highly colour specific, for example DAPI (blue), FITC (green), or Texas red (red) [15, 27, 59]. The same statement holds true if monochrome light sources (lasers) are used. A transformation of the rgb space into a different colour space that separated the intensity is of advantage if "normal" IHC, especially DAB, is applied which results in a brown detection signal [60]. The HSI colour space (hue, saturation, intensity) is an often used transformation of the rgb colour space. It permits an easy computation of colour-specific intensities [44, 59, 61–63].

Image Quality

Images of "good quality" are mandatory to extract content-based information or to derive a certain diagnosis. However, what is good image quality? The answer depends upon the aim to have "good quality", or, in other words, what is meant by good quality. The quality measures of images that serve for a certain diagnosis probably differ from those that are created by artists that are designed to express ideas or personal views of specific events or motions. Thus, we can distinguish between subjective and machine-oriented methods to define image quality. Subjective methods analyse colour and intensity (brightness) adjustment to the individual human perception (morning light, sunset, moonlight, etc.). Machine-oriented methods compare image copies with the original ones, and correct potential differences.

Measurements of digitised light microscopy images are aimed to result in reliable, reproducible, and clinically applicable data [44, 59, 63, 64]. The mandatory image properties include:

1. No vignetting (shading). The grey values of the background pixels should not depend upon their coordinates (spatially independent). In other words, the background grey values of pixels in the x–y gradient image should be zero.
2. The absolute grey value range (0–255 bits) should be spatially independent.

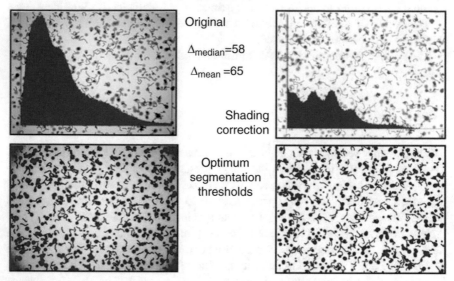

Fig. 4.8 Effect of (automated) image quality correction on segmentation quality

3. The total grey value histogram should display distinct minima and maxima, which indicate the existence of reliable segmentation thresholds.
4. The surface gradient of potential objects should exceed 20 % of the total grey value range (i.e. the objects should be in focus).
5. The image size should exceed 256×256 pixels in order to allow reliable texture analysis.
6. The size of potential objects should amount >1000 pixels/object.
7. The number of potential objects should amount >100 pixels/image.

Some of these requirements can be artificially readjusted. Especially the shading and the grey value range can be corrected to an optimum. An example of automated image quality correction is shown in Fig. 4.8. All these parameters should be continuously measured, corrected, and demonstrated in series as reported by Kayser et al. [65, 66]. Analysis of such series can explain potential inaccurate measurements [16, 59, 65–67].

Object Segmentation

The segmentation and consecutive identification of objects may be difficult especially in IHC images. Objects that are characterised by closed surfaces (nuclei, cells, glands, vessels, etc.) are usually segmented by algorithms that are based upon a dynamic or fixed threshold, and, in addition, upon edge finding mechanisms and analysis of the convexity of the already segmented surface segments [37, 50, 51, 64, 68, 69]. Such algorithms with included variable thresholds and appropriate feedback mechanisms have been reported to reproducibly segment (and identify) more than 95 % of nuclei in IHC images [39]. Membranes are more difficult to segment, especially in IHC images with only partially stained membranes. The identification of nuclei, the computation of the corresponding gravity centres, and the application of Dirichlet's (Voronoi's) tessellation permit the construction of artificial membranes adjusted to the IHC stained membrane fragments [44, 52, 70]. Such an algorithm permits a detailed and reproducible, automated calculation of membrane stains (for example expression of her2_neu) in addition to the crude Her2_neu score [17, 27, 38, 71–73].

Vessels can be segmented in a comparable manner, especially if the inner surface (endothelial cells) is marked by specific antibodies (CD31, CD34, or VEGFR (vascular endothelial growth factor receptor)). The application of appropriate neighbourhood condition and double marker stain (for example Ki67 (MIB2) for proliferation and CD34 for intra-tumour vessels) has been reported to permit a sophisticated insight into the association of tumour cell proliferation and distance from the nearest neighbouring vessel. The situation is shown in Fig. 4.9 and visualises the diffusion density within a malignant tumour [70].

Vascularization and development of lung cancer

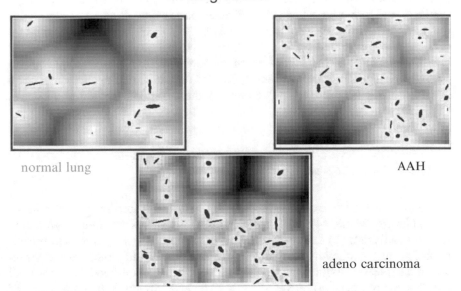

Fig. 4.9 Visualisation of spatial diffusion in relation to vascularisation. Syntactic structure analysis, CD31 IHC, magnification: ×150

Segmentation of FISH signals is relatively easy due to the sharp and demarcated FISH signals. It is, however, handicapped by the small size of the signals and the thin focus plane which might not contain all FISH signals. Therefore, FISH images should be composed by several focus layers in order to possess and analyse all nuclear signals [71, 72].

Analysis of IHC detectable antigens in the cellular cytoplasm is best performed by stereologic methods, such as point/grid counting, in our opinion. An object is the overlay of an appropriate grid (for details, see [74–77]) and the automated assessment of the hit events (points of interest) which are used to calculate stereologic parameters such as area/volume fraction, size adjusted point (object) density, length/area fraction, etc.

Syntactic Structure Algorithms

An object remains the principal and visual controllable main feature of any digitised image. Most objects of microscopic images reflect to nuclei, followed by membranes in images derived from IHC. Once the objects have been segmented, they are considered "as a whole" and their features are measured [36, 45, 47, 56, 58, 78]. There does exist an additional method for analysing and characterising digitised images based upon stochastic geometry, so-called image primitives [44, 49, 79–82]. The idea is that each object or characteristic unit of an image can be constructed by a set of primitives that include {points, lines, circles} above and below a certain threshold [49, 50]. They correspond to peaks (holes) and ridges (valleys), and form a so-called group. Similar to the syntactic structure algorithms that are derived from image objects (usually nuclei), these primitives can serve for extraction of image information. The flexible arrangement of thresholds and the threshold-depending clusters and neighbourhood function is probably an advantage of such algorithms [39]. However, detailed reports of IHC application are still missing to our knowledge. An example of the described method is shown in Fig. 4.10.

Image Information and Feature Extraction

There exist at least five different techniques to derive information from a histological image [37, 38, 44]: (a) analysis of pixel grey values and their coordination (texture analysis); (b) external definition of objects, segmentation, and object-related feature extraction, (c) analysis of object-associated structures and structure-related feature extraction (after definition of a neighbourhood condition), (d) construction of image primitives and related feature extraction, (e) comparison of the individual image "as a whole" with a "set of standardised and quality (content) evaluated set of images (gold standard set)", computation of the best fitting

Example of structural primitives in malignant mesothelioma

Primitives* > threshold

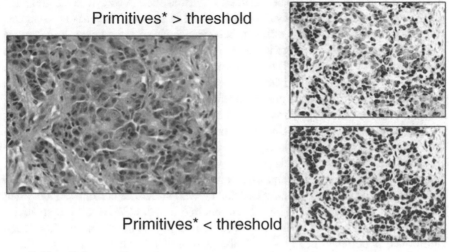

Primitives* < threshold

*) {{<,>}} form a group

Fig. 4.10 Example of evaluation of primitives in a malignant epithelioid mesothelioma. ×180

(nearest neighbouring) image, and association of the (already computed) corresponding labels (diagnosis, texture, objects, structures, and features) with the image under request. Each of the mentioned algorithms has its advantages and disadvantages. They can be considered to be independent from or only weakly dependent upon each other and, therefore, be combined. Especially, the combination of texture, object, and structure features has been applied successfully to automatically screen histological images and to derive crude diagnoses (normal, inflammatory altered, benign and malignant tumours) [37, 50, 51]. The principal features are based upon grey values and their distribution and include size, surface, form factor, moments, pixel-based entropy, moments, transformation-associated features (derived after application of locally dependent and independent filters/transformations such as Fourier, Prewitt, Laplace, Radon, Hadamard, etc.) [57, 82–85]. All these methods result in a vector that can be statistically analysed in respect to observer-related information (diagnosis, prognosis, response to therapy, etc). The applied statistical methods include neural networks and multivariate analysis [27, 44, 56, 63, 80, 86]. The statistical investigations require a training set (to define the diagnosis-associated limitations of the variables, or to train the neural network, and a test set that confirms the calculated results [27, 87–92]. These results can, in addition, be used to search for and select those image areas that contain the "most significant information". What is the most significant image information?

Region of Interest

Already at a first glance it can be stated that a virtual slide (VS) can be disintegrated into image compartments that might be identical in terms of the visualised objects and structures, or that might differ in between. Such an image consists of image compartments that reflect more to the diagnosis compared to others. The areas are called regions of interest (ROI) and can contain an increased frequency of proliferating cells, a denser vascularisation, or less necrotic areas. They are, in addition, considered to be of increased prognostic significant value.

How can these compartments be automatically detected?

All the above discussed techniques can be applied. The image is either divided into connecting compartments equal in size, and the different compartments are analysed in respect of object density and structure parameters (and, in addition, of texture features (sliding technique)), or clusters of specific texture features dependent upon the minimum spanning tree and derived structural entropy are identified [27, 44, 50]. Both techniques have been reported to detect ROIs with a sensitivity >95 %. An example of the obtained ROI in a biopsy of breast (sclerosing ductal carcinoma) is depicted in Fig. 4.11. Other approaches are based upon stereologic sampling and diffusion maps [84, 86, 93, 94], or by density calculations of similarity (dissimilarity) maps [85].

Automated detection of ROI in pleural
biopsy

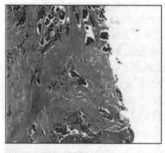

Original

Selected areas each
20% of original

Selected areas by minimum
spanning tree, no relation
to original image size

Fig. 4.11 Example of automated ROI detection in a biopsy of a sclerosing invasive ductal carcinoma of the breast, HE, ×120

Application in IHC

The described methods and algorithms have all been developed to ensure reproducible and reliable measurements of still images and VS. IHC has become the main field of application of this technology in contrast to images acquired from conventional H&E stained slides. Within the IHC application, the most frequently used visualisation technique is based upon the DAB reaction which yields a brown signal in case of antigen presence. The development of automated IHC staining systems ensures a quite stable and reliable reaction despite some problems which are mainly caused by sometime unavoidable tissue preservation and fixation failures. Assuming a high-quality IHC stain, what are the mandatory conditions to perform reproducible staining analysis and quantification?

To answer this question, one should keep in mind the aim of the performed IHC which can be at least twofold: (a) to ascertain/exclude a specific diagnosis (for example in analysing the cytoskeleton), or (b) to measure quantitatively the expression of certain molecules (for example the presence of her2_neu receptors) [11, 12, 16, 20, 67, 71, 76, 87, 88]. Both aims require a different procedure and answer:

(a) The cytoskeleton does consist of a certain subtype (for example keratin 5, 7, 15, etc.) or not. Therefore, the measurement results is "yes/no" despite it may be difficult in rare cases to distinguish between "yes/no" due to a missing or ill-defined threshold. The principal answer can be refined by computing the frequency of tumour cells which contain the specific cytoskeleton; however, this is usually not of major importance. In respect to the clinical request, a "yes/no" result of the applied measurement system is sufficient. Examples include the confirmation/exclusion of metastases and the search for the tumour origin, and the classification of malignant lymphomas (expression of CDs). The distinction between mesothelioma and into the pleura metastasising adenocarcinoma is another example (application of calretinin; see Fig. 4.12). Detailed quantification and syntactic structure analysis of the obtained IHC figures can add information to the aim of IHC and is, however, only seldom performed [50, 52].

(b) The evaluation of the patient's prognosis and of the most adequate therapeutic regimes (for example cytostatic therapy plus specific antibodies in case of lung adenocarcinomas) requires quantitative measurement of the IHC signals. Quantitative measurement means the evaluation of objects expressing positive signals in relation to all objects, which is the sum of positive and negative objects. The number of observed objects should be corrected according to their size if a space (three dimensional) relevant result is aimed. Examples include the evaluation (measurement) of proliferating (Ki67, MIB2 positive) tumour cells which is of prognostic significance in numerous malignancies, and of diagnostic significance in distinguishing benign from malignant tumours (for example typical versus atypical pulmonary carcinoids, threshold 5 %) [55]. Additional examples include membrane stains (her2-neu), or automated analysis of FISH signals [21, 95, 96]. Several commercially available and FDA-approved automated systems serve for these purposes. An automated calculation of the assessed entropy in the IHC investigation is depicted in Fig. 4.12 and that of a

Entropy calculation	H & E	Calretinin
Shannon's entropy:	2.84	4.36
Texture entropy :	6.58	6.79
No of clusters:	23	51
Total entropy (cluster)	14.35	19.94
Entropy Primitives	0.51	0.79
Structure entropy	0.58	0.83

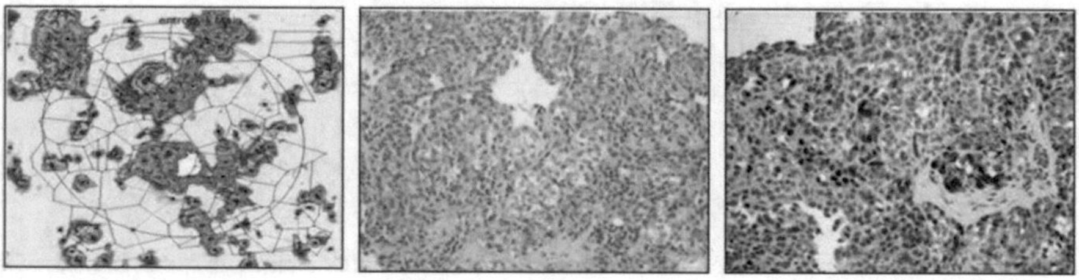

Fig. 4.12 Entropy calculations of HE and IHC (calretinin) stains in a biopsy of a malignant meso-thelioma, ×150

FISH investigation in colon cancer is shown in Fig. 4.13. These examinations are the preceding step towards the so-called predictive diagnosis, and are nowadays followed by gene analysis or investigations of the corresponding protein expression. Examples include the analysis of oncogene pathways, such as Kras, Kraf, Mek, Erk, etc. [73, 97–102]. Until today they are only examined in a yes/no procedure, and, to our knowledge, only a few attempts have been undertaken to get detailed insight into the thermodynamic aspects of such pathways, which might open a new door to cancer therapy.

Perspectives

IHC and related molecular biology and genetic techniques have matured and are nowadays common and routine procedures in institutes of diagnostic surgical pathology. They are embedded in a hierarchical order of tissue-based diagnosis that starts with the examination of conventional H&E stained slides, followed by IHC examinations, and finally analyses of intracellular pathways related to cellular progression and apoptosis. This development of "tissue examinations" is contemporarily accompanied by fast progress of computation and communication, i.e. the digitised world. Image analysis combines both developments as it is based on digitised images provided by whole glass slide scanners, fast standardised connections, and high speed computers. This combination permits the development of fully automated measurements systems, which do not require any sophisticated knowledge of the user [37–39, 51, 65].

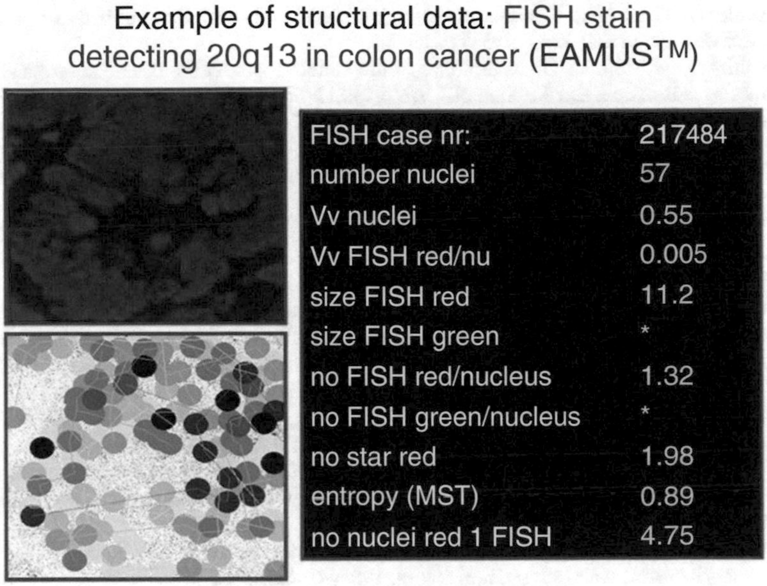

Example of structural data: FISH stain detecting 20q13 in colon cancer (EAMUS™)

FISH case nr:	217484
number nuclei	57
Vv nuclei	0.55
Vv FISH red/nu	0.005
size FISH red	11.2
size FISH green	*
no FISH red/nucleus	1.32
no FISH green/nucleus	*
no star red	1.98
entropy (MST)	0.89
no nuclei red 1 FISH	4.75

Fig. 4.13 Automated evaluation (EAMUSTM, www.diagnomx.eu) of FISH-associated structural data detecting 20q13 alterations in a carcinoma of the colon

A simple formulation of the task (for example measurement of membrane-associated signals) by pressing a corresponding button and some additional information such as image magnification and visualisation dye are already sufficient to perform open access and Internet available IHC measurements [50, 51]. These systems will be matured, and are already able to automatically recognise the applied visualisation stain and the magnification. In addition, the automated recognition of ROIs and corresponding measurements are foreseen in the very near future [27, 44, 52, 85, 86, 93]. The progress in image analysis and digitalisation of a diagnostic pathology laboratory is rapid, and the efforts to integrate such systems into the laboratory work flow are ongoing. IHC will remain as an important chain in tissue-based diagnosis; its quantification will become an additional component of diagnosis assurance and verification in both interactive and automated virtual microscopy.

References

1. Coons A, Creech H, Jones R. Immunological properties of an antibody a containing fluorescent group. Proc Soc Exp Biol. 1941;47:200–2.
2. Kohler G. The Nobel Lectures in Immunology. The Nobel Prize for Physiology or Medicine, 1984. Derivation and diversification of monoclonal antibodies. Scand J Immunol. 1993;37(2): 117–29.
3. Kohler G, Howe SC, Milstein C. Fusion between immunoglobulin-secreting and nonsecreting myeloma cell lines. Eur J Immunol. 1976;6(4):292–5.

4. Kohler G, Milstein C. Continuous cultures of fused cells secreting antibody of predefined specificity. Nature. 1975;256(5517):495–7.
5. Kohler G, Milstein C. Derivation of specific antibody-producing tissue culture and tumor lines by cell fusion. Eur J Immunol. 1976;6(7):511–9.
6. Kohler G, Milstein C. Continuous cultures of fused cells secreting antibody of predefined specificity. 1975. J Immunol. 2005;174(5):2453–5.
7. Taylor C, Burns J. The demonstration of plasma cells and other immunoglobulin containing cells in formalin-fixed, paraffin-embedded tissues using peroxidase labeled antibody. J Clin Pathol. 1974;27:14–20.
8. Blobel GA, et al. The intermediate filament cytoskeleton of malignant mesotheliomas and its diagnostic significance. Am J Pathol. 1985;121(2):235–47.
9. de Jong D, et al. Immunohistochemical prognostic markers in diffuse large B-cell lymphoma: validation of tissue microarray as a prerequisite for broad clinical applications (a study from the Lunenburg Lymphoma Biomarker Consortium). J Clin Pathol. 2009;62(2):128–38.
10. Lin F, Pichard J. Handbook of practical immunohistochemistry. New York: Springer; 2011.
11. Gabius HJ, et al. Reverse lectin histochemistry: design and application of glycoligands for detection of cell and tissue lectins. Histol Histopathol. 1993;8(2):369–83.
12. Gabius HJ, Kayser K. Elucidation of similarities of sugar receptor (lectin) expression of human lung metastases from histogenetically different types of primary tumors. Anticancer Res. 1989;9(6):1599–604.
13. Kayser K, et al. Histopathologic evaluation of application of labeled neoglycoproteins in primary bronchus carcinoma. Hum Pathol. 1989;20(4):352–60.
14. Kayser K, Gabius HJ, Gabius S. Biotinylated ligands for receptor localization. An alternative for immunohistochemistry. Zentralbl Pathol. 1991;137(6):473–8.
15. Gerlach D. Lichtmikroskopie. In: Robenek H, editor. Mikroskopie in Forschung und Praxis. Darmstadt: GIT; 1995.
16. Leong AS. Quantitation in immunohistology: fact or fiction? A discussion of variables that influence results. Appl Immunohistochem Mol Morphol. 2004;12(1):1–7.
17. Albrecht B, et al. Array-based comparative genomic hybridization for the detection of DNA sequence copy number changes in Barrett's adenocarcinoma. J Pathol. 2004;203(3):780–8.
18. Benson M, et al. DNA microarrays to study gene expression in allergic airways. Clin Exp Allergy. 2002;32(2):301–8.
19. Gulmann C, et al. Biopsy of a biopsy: validation of immunoprofiling in gastric cancer biopsy tissue microarrays. Histopathology. 2003;42(1):70–6.
20. Hagemann AR, et al. Tissue-based immune monitoring II: Multiple tumor sites reveal immunologic homogeneity in serous ovarian carcinoma. Cancer Biol Ther. 2011;12(4):367–77.
21. Lin Y, et al. Tissue microarray-based immunohistochemical study can significantly underestimate the expression of HER2 and progesterone receptor in ductal carcinoma in situ of the breast. Biotech Histochem. 2011;86(5):345–50.
22. Warnberg F, et al. Quality aspects of the tissue microarray technique in a population-based cohort with ductal carcinoma in situ of the breast. Histopathology. 2008;53(6):642–9.
23. Hedberg JJ, et al. Micro-array chip analysis of carbonyl-metabolising enzymes in normal, immortalised and malignant human oral keratinocytes. Cell Mol Life Sci. 2001;58(11):1719–26.
24. Yang Z, Tang LH, Klimstra DS. Effect of tumor heterogeneity on the assessment of Ki67 labeling index in well-differentiated neuroendocrine tumors metastatic to the liver: implications for prognostic stratification. Am J Surg Pathol. 2011;35(6):853–60.
25. Sandeck HP, et al. Re-evaluation of histological diagnoses of malignant mesothelioma by immunohistochemistry. Diagn Pathol. 2010;5:47.
26. Yang Q-P, et al. Subtype distribution of lymphomas in Southwest China: analysis of 6,382 cases using WHO classification in a Single Institution. Diagn Pathol. 2011;6:77.
27. Kayser G, et al. Theory and implementation of an electronic, automated measurement system for images obtained from immunohistochemically stained slides. Anal Quant Cytol Histol. 2006;28(1):27–38.

28. Kayser K. [The patho-anatomical Regional Cancer Register of North Baden]. Norm Pathol Anat (Stuttg). 1981;44:1–136.
29. Kayser K, et al. Changes during the last decade in clinical parameters of operated lung carcinoma patients of a center for thoracic surgery and the prognostic significance of TNM, morphometric, cytometric, and glycohistochemical properties. Thorac Cardiovasc Surg. 1997;45(4):196–9.
30. Kayser K, Burkhardt HU. [Incidence of malignant tumors in autopsy records of the Heidelberg Pathological Institute 1900-1975]. Verh Dtsch Ges Pathol. 1978;62:498.
31. Kayser K, Burkhardt HU. The incidence of gastro-intestinal cancer in North Baden (West Germany) 1971–1977. J Cancer Res Clin Oncol. 1979;93(3):301–21.
32. Mireskandari M. How do surgical pathologists evaluate critical diagnoses (critical values)? Diagn Pathol. 2008;3:30.
33. Mireskandari M, et al. Lack of CD117 and rare bcl-2 expression in stomach cancer by immunohistochemistry. An immunohistochemical study with review of the literature. Diagn Pathol. 2006;1:7.
34. Kayser K, Gabius HJ. Graph theory and the entropy concept in histochemistry. Theoretical considerations, application in histopathology and the combination with receptor-specific approaches. Prog Histochem Cytochem. 1997;32(2):1–106.
35. Kayser K, Gabius HJ. The application of thermodynamic principles to histochemical and morphometric tissue research: principles and practical outline with focus on the glycosciences. Cell Tissue Res. 1999;296(3):443–55.
36. Kayser K, Stute H, Tacke M. Minimum spanning tree, integrated optical density and lymph node metastasis in bronchial carcinoma. Anal Cell Pathol. 1993;5(4):225–34.
37. Kayser K, et al. How to measure diagnosis-associated information in virtual slides. Diagn Pathol. 2011;6 Suppl 1:S9.
38. Kayser K, et al. Interactive and automated application of virtual microscopy. Diagn Pathol. 2011;6 Suppl 1:S10.
39. Kayser K, et al. Texture- and object-related automated information analysis in histological still images of various organs. Anal Quant Cytol Histol. 2008,30(6).323–33.
40. Soda M, et al. Identification of the transforming EML4-ALK fusion gene in non-small-cell lung cancer. Nature. 2007;448(7153):561–6.
41. Gutierrez Espeleta GA, et al. BRCA1 and BRCA2 Mutations among Familial Breast Cancer Patients from Costa Rica. Clin Genet. 2012;82(5):484–8.
42. Pan M, et al. Novel LOVD databases for hereditary breast cancer and colorectal cancer genes in the Chinese population. Hum Mutat. 2011;32(12):1335–40.
43. Zhu Q, et al. BRCA1 tumour suppression occurs via heterochromatin-mediated silencing. Nature. 2011;477(7363):179–84.
44. Kayser K, Molnar B, Weinstein RS. Virtual microscopy—fundamentals—applications—perspectives of electronic tissue-based diagnosis. Berlin: VSV; 2006.
45. Kayser K, Kayser G, Metze K. The concept of structural entropy in tissue-based diagnosis. Anal Quant Cytol Histol. 2007;29(5):296–308.
46. Kayser K, Kremer C, Tacke M. Integrated optical density and entropiefluss (current of entropy) in bronchial carcinoma. In Vivo. 1993;7(4):387–91.
47. Kayser K, et al. Combined morphometrical and syntactic structure analysis as tools for histomorphological insight into human lung carcinoma growth. Anal Cell Pathol. 1990;2(3):167–78.
48. Kayser K, Borkenfeld S, Goldmann T, Kayser, G. To be at the right place at the right time. Diagn Pathol. 2011;6:68. doi:10.1186/1746-1596-6-68.
49. Kayser K. Quantification of virtual slides: approaches to analysis of content-based image information. J Pathol Inform. 2011;2:2.
50. Kayser K, et al. AI (artificial intelligence) in histopathology—from image analysis to automated diagnosis. Folia Histochem Cytobiol. 2009;47(3):355–61.
51. Kayser K, et al. Grid computing in image analysis. Diagn Pathol. 2011;6 Suppl 1:S12.
52. Kayser K, et al. Theory of sampling and its application in tissue based diagnosis. Diagn Pathol. 2009;4:6.

53. Voronoi G. Nouvelles applications des parametres continus a la theorie des formes quadratiques, deuxieme memoire: recherches sur les paralleloedres primitifs. J Reine Angew Math. 1902;134:188–287.
54. Brinkhuis M, et al. Minimum spanning tree analysis in advanced ovarian carcinoma. An investigation of sampling methods, reproducibility and correlation with histologic grade. Anal Quant Cytol Histol. 1997;19(3):194–201.
55. Kayser K, et al. Carcinoid tumors of the lung: immuno- and ligandohistochemistry, analysis of integrated optical density, syntactic structure analysis, clinical data, and prognosis of patients treated surgically. J Surg Oncol. 1996;63(2):99–106.
56. Meijer GA, et al. Syntactic structure analysis of the arrangement of nuclei in dysplastic epithelium of colorectal adenomatous polyps. Anal Quant Cytol Histol. 1992;14(6):491–8.
57. Prewitt JMS, Wu SC. An application of pattern recognition to epithelial tissues. In: Computer applications in medical care. IEEE Computer Society; 1978.
58. van Diest PJ, et al. Syntactic structure analysis. Pathologica. 1995;87(3):255–62.
59. Leong FJ, Leong AS. Digital imaging in pathology: theoretical and practical considerations, and applications. Pathology. 2004;36(3):234–41.
60. Pham NA, et al. Quantitative image analysis of immunohistochemical stains using a CMYK color model. Diagn Pathol. 2007;2:8.
61. Leong FJ, Brady M, McGee JO. Correction of uneven illumination (vignetting) in digital microscopy images. J Clin Pathol. 2003;56(8):619–21.
62. Leong FJ, McGee JO. Automated complete slide digitization: a medium for simultaneous viewing by multiple pathologists. J Pathol. 2001;195(4):508–14.
63. Leong FJ, McGee JO. An automated diagnostic system for tubular carcinoma of the breast—an overview of approach and considerations. Stud Health Technol Inform. 2003;97:57–72.
64. Leong FJ, Leong AS. Digital imaging applications in anatomic pathology. Adv Anat Pathol. 2003;10(2):88–95.
65. Kayser K, et al. Image standards in tissue-based diagnosis (diagnostic surgical pathology). Diagn Pathol. 2008;3:17.
66. Kayser K, et al. How to measure image quality in tissue-based diagnosis (diagnostic surgical pathology). Diagn Pathol. 2008;3 Suppl 1:S11.
67. Leong AS. Pitfalls in diagnostic immunohistology. Adv Anat Pathol. 2004;11(2):86–93.
68. Chiverton J, Xie X, Mirmehdi M. Automatic bootstrapping and tracking of object contours. IEEE Trans Image Process. 2012;21(3):1231–45.
69. Mishra AK, Aloimonos Y, Cheong LF, Kassim AA. Active visual segmentation. IEEE Trans Pattern Anal Mach Intell. 2012 ;34(4):639–53. doi:10.1109/TPAMI.2011.171.
70. Kayser K, et al. Atypical adenomatous hyperplasia of lung: its incidence and analysis of clinical, glycohistochemical and structural features including newly defined growth regulators and vascularization. Lung Cancer. 2003;42(2):171–82.
71. Atkinson R, et al. Effects of the change in cutoff values for human epidermal growth factor receptor 2 status by immunohistochemistry and fluorescence in situ hybridization: a study comparing conventional brightfield microscopy, image analysis-assisted microscopy, and interobserver variation. Arch Pathol Lab Med. 2011;135(8):1010–6.
72. Reljin B, et al. Breast cancer evaluation by fluorescent dot detection using combined mathematical morphology and multifractal techniques. Diagn Pathol. 2011;6 Suppl 1:S21.
73. Pradeep CR, et al. Modeling ductal carcinoma in situ: a HER2-Notch3 collaboration enables luminal filling. Oncogene. 2012;31(7):907–17.
74. Gundersen HJ. Stereology of arbitrary particles. A review of unbiased number and size estimators and the presentation of some new ones, in memory of William R. Thompson. J Microsc. 1986;143(Pt 1):3–45.
75. Gundersen HJ, Andersen BS, Floe H. Estimation of section thickness unbiased by cutting-deformation. J Microsc. 1983;131(Pt 1):RP3–4.
76. Gundersen HJ, Jensen TB, Osterby R. Distribution of membrane thickness determined by lineal analysis. J Microsc. 1978;113(1):27–43.
77. Gundersen HJ, Osterby R. Optimizing sampling efficiency of stereological studies in biology: or 'do more less well!'. J Microsc. 1981;121(Pt 1):65–73.

78. Unverzagt C, et al. Structure-activity profiles of complex biantennary glycans with core fucosylation and with/without additional alpha 2,3/alpha 2,6 sialylation: synthesis of neoglycoproteins and their properties in lectin assays, cell binding, and organ uptake. J Med Chem. 2002;45(2):478–91.

79. Kayser K, et al. Application of attributed graphs in diagnostic pathology. Anal Quant Cytol Histol. 1996;18(4):286–92.

80. Villacorta JA, et al. Mathematical foundations of the dendritic growth models. J Math Biol. 2007;55(5-6):817–59.

81. Mattfeldt T, et al. Classification of spatial textures in benign and cancerous glandular tissues by stereology and stochastic geometry using artificial neural networks. J Microsc. 2000;198(Pt 2):143–58.

82. Stoyan D, Kendall WS, Mecke J. Stochastic geometry and its applications. Berlin: Akademie; 1987.

83. Voss K, Süße H. Praktische Bildverarbeitung. München: Carl Hanser; 1991.

84. Oger M, et al. Automated region of interest retrieval and classification using spectral analysis. Diagn Pathol. 2008;3 Suppl 1:S17.

85. Romo D, Romero E, Gonzalez F. Learning regions of interest from low level maps in virtual microscopy. Diagn Pathol. 2011;6 Suppl 1:S22.

86. Oger M, Belhomme P, Gurcan MN. Classification of low resolution virtual slides from breast tumor sections: comparison between global and local analysis. Conf Proc IEEE Eng Med Biol Soc. 2009;2009:6671–4.

87. Bartels P, et al. Structural and biophysical determinants of single Ca(V)3.1 and Ca(V)3.2 T-type calcium channel inhibition by N(2)O. Cell Calcium. 2009;46(4):293–302.

88. Bartels PH, Bartels HG. Discriminant analysis. Anal Quant Cytol Histol. 2009;31(5):247–54.

89. Bartels PH, Montironi R. Quantitative histopathology: the evolution of a scientific field. Anal Quant Cytol Histol. 2009;31(1):1–4.

90. Bartels PH, et al. Carcinogenesis and the hypothesis of phylogenetic reversion. Anal Quant Cytol Histol. 2006;28(5):243–52.

91. Kristensen S, Mainz J, Bartels P. Selection of indicators for continuous monitoring of patient safety: recommendations of the project 'safety improvement for patients in Europe'. Int J Qual Health Care. 2009;21(3):169–75.

92. Montironi R, et al. Decision support systems for morphology-based diagnosis and prognosis of prostate neoplasms: a methodological approach. Cancer. 2009;115(13 Suppl):3068–77.

93. Belhomme P, et al. Generalized region growing operator with optimal scanning: application to segmentation of breast cancer images. J Microsc. 1997;186(Pt 1):41–50.

94. Belhomme P, et al. Towards a computer aided diagnosis system dedicated to virtual microscopy based on stereology sampling and diffusion maps. Diagn Pathol. 2011;6 Suppl 1:S3.

95. Kayser G, et al. Numerical and structural centrosome aberrations are an early and stable event in the adenoma-carcinoma sequence of colorectal carcinomas. Virchows Arch. 2005;447(1):61–5.

96. Szoke T, et al. Prognostic significance of endogenous adhesion/growth-regulatory lectins in lung cancer. Oncology. 2005;69(2):167–74.

97. Benetatos L, Vartholomatos G. Deregulated microRNAs in multiple myeloma. Cancer. 2012;118(4):878–87.

98. Flaherty KT, Fisher DE. New strategies in metastatic melanoma: oncogene-defined taxonomy leads to therapeutic advances. Clin Cancer Res. 2011;17(15):4922–8.

99. Paulin R, et al. From oncoproteins/tumor suppressors to microRNAs, the newest therapeutic targets for pulmonary arterial hypertension. J Mol Med (Berl). 2011;89:1089.

100. Puzio-Kuter AM. The role of p53 in metabolic regulation. Genes Cancer. 2011;2(4):385–91.

101. Watson AJ, Collins PD. Colon cancer: a civilization disorder. Dig Dis. 2011;29(2):222–8.

102. Zheng HC, Takano Y. NNK-induced lung tumors: a review of animal model. J Oncol. 2011;2011:635379.

Chapter 5
Immunohistochemistry in the Study of Cancer Biomarkers for Oncology Drug Development

Fang Jiang and Evelyn M. McKeegan

Introduction

The immunohistochemistry (IHC) assay is based on the binding of antibodies (Ab) to specific antigens (Ag) (protein–protein interaction) in tissue sections or cytological preparation slides, and it is visualized using conventional microscopy [1–3]. IHC assays provide valuable information for basic research, and for preclinical studies through to clinical trials to understand the mechanisms of action of compounds in drug development. Biomarkers—characterized as a specific underlying pathophysiological state that can be objectively measured and evaluated to detect or define disease progression, or predict or quantify therapeutic responses [4]—identified by IHC assays are essential to many diagnoses in histopathology directly affecting treatment and patient care. The assays may be used for companion diagnostic purposes and can influence critical clinical decision making [2, 5–8].

In cancer research, IHC is essential to identify the cell type from which the cancer originated [1]. Predictive biomarker studies, including those involving tumor tissues, body fluid proteins, blood cells, and genotype, have the power to provide diagnostic, therapeutic, and prognostic information for personalized medicine [9]. The most relevant biomarkers identified using IHC assay are at the tumor tissue level, which allows the applications of histogenic biomarkers and target proteins to assist cancer therapy [10–12] and stratified medicine [13–15]. Tumor tissues can be

F. Jiang (✉) • E.M. McKeegan
Cancer Discovery, Global Pharmaceutical Research and Development, Abbott Laboratories, AP10, 100 Abbott Park Road, Abbott Park, IL 60064, USA
e-mail: fang.jiang@abbott.com

© Springer Science+Business Media New York 2016
S.A. Aziz, R. Mehta (eds.), *Technical Aspects of Toxicological Immunohistochemistry*, DOI 10.1007/978-1-4939-1516-3_5

classified using antibodies to receptors, cytoskeletal components, enzymes, blood group antigens, and specific structures that serve as markers for particular cell types or tissues. Treatment-associated increases or decreases in proteins as measured by IHC can provide insights into physiological responses. For example, angiogenesis as a hallmark of solid tumors is critical to the transition of premalignant lesions in a hyperproliferative state to the malignant phenotype, which leads to tumor growth and metastasis. The intensity of angiogenesis, assessed by counting microvessels in neoplastic tissue, acts as a prognostic factor for many solid tumors such as breast, prostate, glioma, ovarian, and lung [16–20]. When tumor samples are available, antiangiogenic therapy can be assessed using baseline or serial measurements of gene and protein expression, microvessel density, perivascular cell coverage, interstitial fluid pressure, intratumor oxygen tension, and drug uptake [9]. Analysis of cell suspensions from blood, bone marrow, or other sources by immunocytochemistry (ICC) is also commonly used to identify biomarkers for drug development. For instance, enumeration of circulating tumor cells (CTC) is a clinical test and can accompany clinical trials for new drug development [4, 6].

This chapter discusses technical aspects of IHC and a basic procedure: development of a multi-fluorophore labeling technique for tumor vascular identification in preclinical human xenograft tumor models. The chapter includes a study demonstrating the use of IHC assays to evaluate tumor vascular biomarkers in antiangiogenic drug development using linifanib (ABT-869, a multitargeted receptor tyrosine kinase (RTK) inhibitor) as an example, including measurement of the drug-target proteins phospho-vascular endothelial growth factor receptor 2 (pVEGFR 2) and phospho-platelet-derived growth factor receptor β (pPDGFR β), microvessel density and diameter, pericyte coverage as an indicator of vascular maturation, and hypoxic protein expression. The chapter also discusses the use of multi-fluorophore labeling technique to detect rare CTC from patient blood samples for oncology drug development.

Technical Aspects of IHC

The fundamental concept of IHC is the detection of antigen in tissue sections by antibody binding. Once antigen–antibody binding occurs, it is visualized by a histochemical chromagen reaction or by fluorochromes visible by conventional light microscopy or fluorescent microscopy [3]. The assays are performed on tissue sections or cytological preparations placed onto microscope slides. Multiple target proteins may be detected on the same specimen using different antibodies on the single tissue slide or serial sections of a tissue sample (Fig. 5.1a–c). Alternately, a single target can be detected on multiple tissues on a single slide with multiple tissue cores. These tissue microarrays (TMA) allow the measurement target expression in hundreds of tumor samples on the same slide (Fig. 5.1d) [21, 22].

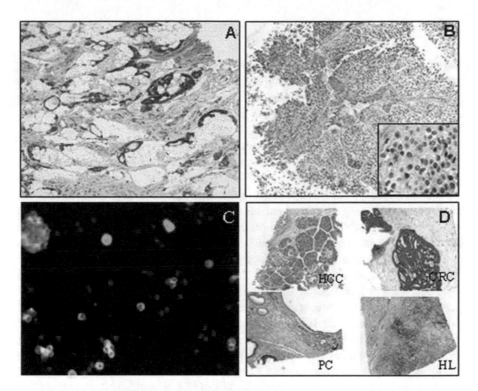

Fig. 5.1 Identification of target proteins using IHC on tissue sections or cytological preparation slides. HRP/DAB detection method was used for all images except C. (**a**) Endothelin A expression on human prostate cancer tissue. (**b**) Phospho-c-Myc expression in the nuclei of human kidney carcinoma tissue. *Insert*: enlarged image. (**c**) Multi-fluorophore detection of target proteins on HCT-15 human colon cancer cells, spiked into healthy subject PBMC. Tumor cell heterogeneity is visible. *Green*—Alexa 488 conjugated anti-cytokeratin 8/18 (CK8/18) monoclonal Ab; *red*—Alexa 594 conjugated anti-epithelial platelet cell adhesion molecule (EpCAM) monoclonal Ab; *orange*—Alexa 555 conjugated anti-breast cancer 1 (BRCA1) polyclonal antibody; *blue*—DAPI. (**d**) Human cancer tissue microarray stained with polyclonal anti-Bcl-XL antibody. *HCC* hepatocellular carcinoma, *CRC* colon rectal carcinoma, *PC* prostate cancer, *HL* Hodgkin's lymphoma

Basic Procedure of IHC

IHC is typically performed on 4–6 μm thick formalin-fixed paraffin-embedded (FFPE) tissue slices or a justified thickness (8–90 μm) of frozen fresh tissue. Through a series of incubation steps the primary antibody specifically binds the target protein and is visualized using an enzyme-linked chromagen or fluorophore conjugate.

Fixation

To adequately preserve cellular proteins and avoid tissue autolysis after sample collection, tissues need to be fixed [3, 23]. Two types of fixatives are used in histopathology: cross-linking (non-coagulating) fixatives and coagulating fixatives. A commonly used cross-link fixative is formaldehyde, which preserves mainly peptides and the general structure of cellular organelles for histopathological assessment and IHC interpretation. The mechanism of fixation with formaldehyde is the formation of cross-links between the formalin and uncharged reactive amino groups (–NH or NH_2). The cross-links result in a profound change in the conformation of macromolecules and allow recognition of proteins (Ag) by antibodies [24]. Formaldehyde fixation is a progressive time- and temperature-dependent process. Overfixation commonly occurs during the FFPE process and can produce false-negative results from excessive cross-links [23]. Therefore, an Ag retrieval procedure to restore the protein (Ag) configuration is commonly used to avoid false-negative results for FFPE tissue sections. In our tumor vascular study, the tumor samples were processed by either the FFPE process or the frozen fresh tissue process. For FFPE, the 5 mm^3 tumor samples were collected and immediately immersed in 10 % neutral formalin for at least 2 h prior to the FFPE process. For the frozen procedure, 5 mm^3 to 1 cm^3 fresh tumor samples were collected and immediately snap frozen by immersing in liquid nitrogen. Samples were kept for up to 2 h in liquid nitrogen as samples were collected, and then all samples were transferred to an −80 °C freezer. Cryosections were generated from a −20 °C cryostat and fixed in 4 % paraformaldehyde, pH 7.4, for 1 h, washed in PBS, air dried, and stored at 4 °C.

For the circulating tumor cell study, a light formaldehyde fixation (1 %, for 5 min) was used for cytological slide preparation from fresh samples, followed by a PBS buffer rinse and air dry. Slides were stored at −80 °C with desiccants for up to 1 year.

FFPE Tissue Section and Frozen Fresh Tissue Cryosection

IHC assays are performed on tissues in various forms, including FFPE tissues, frozen fresh tissues, and cytological preparations or smears. The choice between the various preparations depends on research needs for preservation of cell/tissue morphology and duration of storage. FFPE tissue blocks provide optimal morphology and are easy to handle and store. For general morphological assessment and IHC protein evaluation, 4–6 μm FFPE sections are commonly used. However, the antigenicity may decrease with longer storage times; the Ag retrieval procedure can help to restore the antigenicity. Cryosections provide better antigenicity of proteins with rapid turnover times but require quick procedures such as snap frozen tissues, use of a low temperature cryostat for sectioning, and immediate fixation. If a frozen tissue is not sectioned carefully or the tissue blocks are stored for long durations, poor tissue morphology may result. In our vascular study, for the purpose of generating data rapidly to accommodate drug discovery timelines, we used the frozen fresh tumor process for most of the tumor vascular studies. 90 μm thickness cryosections were obtained for global tumor vascular observation and 20 μm thickness cryosections, recommended for better length of microvessel observation, were obtained for the identification of target proteins such as the phosphorylated receptors for PDGF and VEGF, microvessels, pericytes, and hypoxic proteins. We particularly chose 20 μm thicknesses for the study because the vasculature presented good morphology not only in the length of the vessel but also pericyte coverage, which provided additional information for the development of linifanib [25, 26].

Tissue Pretreatment and Antigen Retrieval

The effects of formaldehyde can be partially reversed [27] to restore antigenic epitopes of the target protein prior to antibody binding [1, 28]. In general, several different tissue pretreatments for FFPE tissue sections should be tested to determine which approach provides the best access of the antibody to the antigen epitopes. Pretreatment options include heat-induced epitope retrieval (HIER) [29] and enzymatic treatment with proteolytic enzymes, which presumably breaks the formalin-induced methylene cross-links in the antigenic molecules to restore their immunoreactivity. In the cancer biomarker studies, to unmask antigenicity of pPDGFR β, pVEGFR 2, von Willebrand factor (vWF), C-myc, endothelin A (ETA), and Bcl-xl proteins from FFPE tissue sections, the pretreatment procedure was performed after tissue sections were dewaxed and rehydrated by a graded alcohols procedure. The tissue sections were heated in the presence of Tris buffer pH 9 solutions as a Target Retrieval Solution using a pressure cooker at 120 °C for 2 min (Digital Programmable Pressure Cooker, Biocare Medical, Walnut Creek, CA). In the absence of pretreatment, the target proteins could not be detected or were poorly detected by the respective antibody, demonstrating the need in this case of using an antigen retrieval procedure (Fig. 5.2). There would be no need to use this procedure with frozen fresh tissue cryosections.

Blocking Considerations

Protein blocking The blocking solution contains nonspecific proteins isolated from the species in which the antibody was made to prevent nonspecific binding of the primary or secondary antibodies. The blocking reagent can also comprise serum-free recombinant blocking proteins, and is used universally for any detection system. The blocking step usually includes an incubation period for approximately 5–10 min prior to the antibody incubation. We used a serum-free blocking reagent for 5 min before the primary Ab incubation and the results were satisfactory (Figs. 5.1 and 5.2).

Enzyme blocking Endogenous peroxidase in blood cells can react with DAB (3′, 3′-diaminobenzidine) to produce a brown product indistinguishable from specific

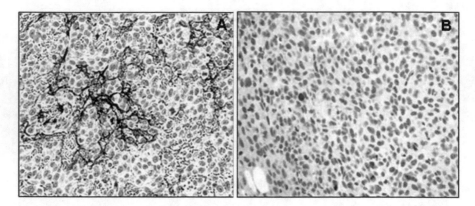

Fig. 5.2 vWF antibody immunostaining on the HT1080 human xenograft fibrosarcoma tissue to detect endothelial cells. (**a**) IHC staining without antigen retrieval. No vessels are visible. (**b**) IHC staining after antigen retrieval (HIER). Characteristic tumor vasculature including dilated, saccular, and disorganized vessels is notable. HRP/DAB detection method. Hematoxylin counterstain

immunostaining using the HRP/DAB (horseradish peroxidase/DAB) detection system [1]. The native peroxidase activity in tissue sections can be destroyed almost completely using a diluted solution of hydrogen peroxide (0.03–3 % H_2O_2). In our tumor vascular IHC studies, 3 % hydrogen peroxide was used to ensure the xenograft tumor tissues, which had abundant hemorrhages, had clean backgrounds. The tissue sections were incubated in 3 % hydrogen peroxide for 5 min before the primary antibody incubation, and buffer rinses were performed after primary antibody incubation. The buffer washes must be performed before the HRP incubation.

There is another commonly used chromagenic detection system, alkaline phosphatase/fast red, in which 2–3 % levamisol is used to block endogenous alkaline phosphatase.

Fc blocking Fc receptors of mononuclear blood cells are usually destroyed during FFPE procedure but can cause nonspecific binding of the primary antibody to lightly fixed frozen tissues [1, 3]. During the tumor vascular study, we used 4 % paraformaldehyde to fix 20 μm frozen sections for an hour and stained with individual antibodies that were directly conjugated with fluorophores, including pPDGFR β, CD31, α-SMA, and collagen IV. The background was at an acceptable level without Fc blocking step (Fig. 5.3).

Primary Antibody and Controls

In IHC, the primary Ab is used to bind and recognize a specific antigen in tissue sections or cytology slides. Primary Abs can be monoclonal or polyclonal and made in diverse species. One requirement for optimization of an IHC assay is to establish

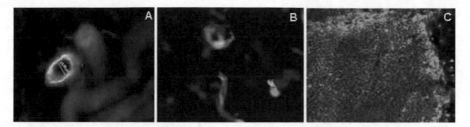

Fig. 5.3 Multi-fluorophore labeling used in pericyte coverage and microvessel measurements. 20 μm cryosections stained with various antibodies in the absence of an Fc blocking step. (**a**) Dual labeled microvessel and pericyte on HT1080 human fibrosarcoma tumor-bearing mouse skeletal muscle. Lectin/FITC perfusion for detection of vessels (*green*) and polyclonal Ab α-SMA/Alexa 594 staining for pericytes (*red*). (**b**) Dual labeling of vessel and basement membrane on normal rat brain tissue. Lectin/FITC perfusion for detection of vessels (*green*) and polyclonal Ab collagen IV/Alexa 594 staining for structural protein expression in matrix around vessels. (**c**) Multi-fluorophore labeling of SW620 human colon tumor tissue. Drug distribution assessed by i.p. administration of Paclitaxel/Oregon green to tumor-bearing mouse. Hypoxic protein expression was detected by mIgG/Alexa 647 conjugate (*yellow*). Lectin/Alexa 594 perfusion for microvessels. *Blue*—nuclei/DAPI

that the antibody binds only to the protein containing the immunogen peptide. Although each Ab is generated against a particular immunogen, dilution of the primary antibody still needs to be optimized with appropriately chosen controls. Controls for the specificity of an antibody are important for the interpretation of protein localization and expression [30]. Specific controls for assay development comprise positive and negative controls. The positive controls use cells or tissues known to contain the protein. The negative controls use cells or tissues known not to contain the protein, or replacing the primary antibody with serum or an isotype control antibody and lastly a preabsorption control, in which the antibody is mixed with the protein or peptide used to generate the antibody. Primary antibody incubation times and temperatures range from an hour at room temperature or 37 °C to overnight at 4 °C. The working concentration of primary Ab for IHC assay generally ranges from 2 to 10 μg/mL. After incubation is completed, the tissue sections should be rinsed thoroughly to remove excess Ab.

In linifanib development, the Abs we used for vascular biomarker identification, including pPDGFR β, pVEGFR 2, vWF, CD31, α-smooth muscle actin (α-SMA), collagen IV, and hypoxic protein, are commercially available. Each antibody was generated against synthetic peptides and the specificities determined by immunoblot and/or immunoprecipitation, greatly reducing the possibility of binding to epitopes not found on the original peptides [25, 26, 31]. The primary antibodies were incubated with tissues for 2 h at room temperature with gentle horizontal rotation or stored at 4 °C overnight. Matched isotype control antibodies for all primary antibodies and preabsorption controls for pPDGFR β and pVEGFR 2 were used.

Detection Systems

Detection systems are classified as direct and indirect. Each method uses a label conjugated to the primary Ab, secondary Ab, or tertiary Ab to allow visualization of the tissue Ag–Ab immune complexes with microscopy. The detection system needs to be carefully selected during assay development taking into consideration the Ab specificity, color discrimination, cell type, and tissue origin. A variety of labels have been used, including fluorescent compounds, enzymes, and metals [32].

Direct Methods

Direct methods of detection are straightforward, based on the conjugation of the primary antibody to a label. Different labels that have been used include fluorochromes, enzymes, colloidal gold, and biotin [3]. With direct methods, the prepared tissue sample is incubated with antibody-label conjugate. Examples of fluochrome labeling are the Alexa dyes for fluorescence detection. One indirect method but referred to as a specific method is the Zenon labeling system (Invitrogen, Grand Island, NY). Zenon labeling technology utilizes a fluorophore-, biotin-, or enzyme-labeled Fab fragment directed against the Fc domain of the primary antibody. Fab fragments specific for human or mouse IgG1, IgG2a, IGg2b, and rabbit IgG are available. The labeling method does not require purification of the labeled antibody and intra-species cross reactivity is low. Since the Fab fragment is provided as a fluorophore conjugate, we were able to have quick turnaround times for assay development, avoid nonspecific antibody binding, and obtain sufficient signals for imaging data analysis for the tissue-based tumor vascular study and the cell-based CTC study. Moreover, this technology permits the use of multiple antibodies in a single staining procedure. Each antibody is labeled with a different.

Indirect Method

An indirect system increases the sensitivity of detection by introducing a labeled secondary antibody which recognizes the primary antibody. If the primary Ab is made in rabbit, the secondary should bind to the rabbit Ab. The secondary antibody is labeled with an enzyme, such as HRP or fluorescent dye, includes a polymer that increases the sensitivity of the detection system, and reduces the background staining [33]. Multiple molecules of enzyme and the secondary antibody are attached to this polymer, which simplifies the three-step methods such as ABC (avidin–biotin–peroxidase complex) while maintaining equal or higher sensitivity.

The sensitivity of this method is higher than a direct method due to signal amplification: the increased ratio of label to primary antibody increases the intensity of the reaction [3]. However, specificity may be lower because of nonspecific binding by the secondary Ab. We used a commercially available polymer-based labeling two-step method in the HRP/DAB chromagenic detection system to detect diverse target proteins (Fig. 5.1). The incubation time was 20–30 min and PBS buffer was used to rinse prior to the use of a chromagen.

Combination of Detection Systems for Multiple Target Protein Detection

Different IHC detection methods can be incorporated into a single protocol to simultaneously visualize multiple components. For example, antibody-based detection of antigen epitopes can be combined with non-antibody-based methods such as lectin binding to glycoproteins on surface of endothelial cell. One such protocol is detailed in section "Assay Development".

Chromagens

DAB is often used as chromagen for HRP-labeled antibodies. DAB reacts with the peroxidase enzyme in the presence of hydrogen peroxide and is converted to a dark brown insoluble precipitate which provides high resolution. The incubation time in DAB is 3–5 min, while the negative control is checked to obtain the necessary high quality of resolution. Buffer rinses are performed after incubation is complete. Fast red is a commonly used dye for the alkaline phosphatase (ALP) detection system. Fluorescent labeled secondary antibodies can also be used to boost signals.

Counterstaining

Counterstaining is the final step in IHC. It enhances the morphology and contrast of cells and tissues. Hematoxylin is the most commonly used counterstain in chromagenic detection methods. The molecule binds to the acidic components in nuclei, resulting in a purplish-blue stain. The duration of exposure to hematoxylin is approximately 30 s to 1 min, based on the color appearance. 0.8 % ammonium chloride or PBS buffer rinse can speed the nuclei blue process. The sample is rinsed with water after the process is complete. Slides can be air dried prior to placement of a coverslip.

Fluorescence stained slides are washed with PBS, rinsed with distilled H_2O, air dried, and covered with prolong anti-fade mounting medium containing DAPI (4', 6-diamidino-2-phenyindole, Invitrogen), which immediately stains the nuclear DNA fragments.

Coverslip

When the IHC process is complete, the stained slides should be covered by mounting medium for microscopic examination and storage. Several applications are commercially available to meet the purpose. Usually, chromagen stained slides are dehydrated through a series of increasing alcohol concentrations, rinsed in xylenes, and coverslipped in permanent mounting medium for long storage. In our studies, we used aqueous mounting medium (Sigma, St. Louis, MO) to mount both tissue slides and cell slides directly after air drying. Similar results were obtained without dehydration using alcohol. For fluorophore detection, a mounting medium containing DAPI as described above was used to coverslip after slides were air dried. The results were satisfactory.

IHC for Cancer Biomarker Studies

Cancer biomarker studies using IHC for target protein identification can answer three questions: can we identify useful targets by differential cell expression, is our target present on the cells of interest, and is our protein of interest modified by treatment. Our tumor vascular study and circulating tumor cell study elaborated the utilization of IHC in oncology drug development.

Cancer Biomarker Studies on Tissue Section Slides

IHC localizes a target protein at the microscopic level, which allows us to better understand the cellular expression pattern in normal and disease tissue as well as the subcellular localization pattern. In oncology drug early development programs, tests are generally performed in xenograft tumor models to evaluate whether particular target proteins may be present as biomarkers in clinical samples. The preclinical studies provide baseline information for the therapeutic effect in tumor cells. With IHC, the cell-specific expression pattern of the marker within the tumor can be determined; results from such studies have been used to make go/no go decisions [34]. For oncology drug development at Abbott (AbbVie), IHC assays provide cell-specific detection of the target protein and information on what cell type expresses the target. The specific target protein expression, or change of expression level and localization, has helped us to understand the biological response to oncology drug candidates (Figs. 5.1 and 5.4). Particularly in linifanib development, the IHC studies on target proteins phospho-VEGFR 2, which is expressed predominantly in endothelial cells and tumor cells, and phospho-PDGFR β, which is expressed primarily in pericytes and tumor cells, in xenograft tumor models of HT1080 (human fibrosarcoma), SW620 (human colon carcinoma), and

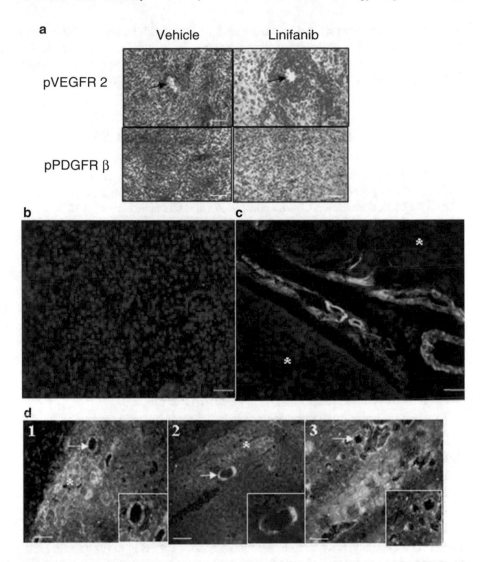

Fig. 5.4 Tumor vascular identification in antiangiogenic drug development. (**a**) pPDGFR β and pVEGFR 2 immunostaining on 9L rat xenograft glioma tissue using HRP/DAB detection method. Linifanib inhibited phosphorylation of both receptors on endothelial cells and tumor cells. *Arrows* indicate microvessels. (**b**) U87-MG human xenograft glioma tumor tissue stained with CD31/ Alexa 594 stained for microvessels (*red*). (**c**) Dual fluorophore labeling of vessels and pericytes in HT1080 fibrosarcoma tissue. Endothelial cells (*red*) form vessels from the host tissue into the tumor; pericyte (*green*) coverage follows. Lectin/Alexa 594 vascular perfusion for vessels (*red*) and α-SMA/Alexa 488 staining for pericyte coverage (*green*). *Asterisks*: Tumor islet. (**d**) Dual labeling of α-SMA/FITC (*green*) and pPDGFR β/Alexa 594 (*red*) on HT1080 fibrosarcoma tissue demonstrated co-localization of the two proteins. (**d1**) No co-localization was detectable on a vessel located in host muscle tissue. (**d2**) Some co-localization is visible on a vessel located between host and tumor tissue. (**d3**) Co-localization presented on vessels in tumor area. *Asterisks*: Host tissue. *Arrows* point vessels. *Blue* indicates nuclei for all images. *Bars* represent 50 μm in **a**, **b**, and **d**, 20 μm in **c**

9L (rat glioma), demonstrated that treatment with linifanib strongly inhibited the immunostaining with both phosphorylation-specific antibodies to the receptors. From this data, we concluded that inhibition of the primary angiogenic targets in vivo (Fig. 5.4a) correlated with tumor growth inhibition [25, 26].

Cancer Biomarker Study on Cytological Preparation Slides

Immunocytochemistry (ICC) is frequently used to characterize circulating tumor cells (CTC) as potential biomarkers. After the CTC population is enriched and placed on slide, data derived from enumeration and characterization of CTC using ICC have the potential to stratify cancer patients for outcome and to provide information on patient response to therapy [35–37] Sensitive and reproducible ICC assays are required to identify the patients most likely to benefit from the treatment. Accurate results from ICC are challenging to obtain, particularly for CTC due to their low abundance in blood and because multiple cell surface markers are ordinarily required to correctly identify a CTC. Our CTC study combined the classification of immunostaining pattern in diverse blood nucleated cells with cell morphological assessment. To detect and enumerate CTC, multi-fluorophore labeling is required to exclude non-CTC cell populations, mostly leukocytes, and identify tumor marker(s) expressed on CTC (Fig. 5.1c). Since CTC are rare in blood, ranging from zero to thousands per mL of blood, an automated scanning system for immunostained slides and software for image analysis are also required [38, 39].

Assay Development

General Assay Development and Validation

To answer questions with IHC, it is important to properly develop and optimize the assays for each study. Several steps are recommended for assay development. First, western blot or immunoprecipitate is used to determine each antibody's specificity and, subsequently, assess several different antibodies ensuring that the antibody titer and incubation times are optimized to find a suitable Ab to the target protein. The goal is to demonstrate similar IHC reactivity with at least two different Abs generated against the same target. Another factor to consider is that not all Abs react well on tumor tissues. Second, to optimize antibody binding to its target epitope, determine if the tissue needs antigen retrieval (most of FFPE tissue) with appropriate processes. Frozen fresh tissue will not require antigen retrieval. Pretreatment or Ab retrieval techniques can make the difference between high quality of IHC staining and no staining at all (Fig. 5.2). Third, prior to testing target tissues, carefully select cell lines or tissues for positive and negative controls. After the controls are tested, a panel of tissue samples, i.e., tissue microarray slides, needs to be tested to calibrate the staining pattern and intensity. To obtain uniform results in which data can be compared across experiments, the following are usually necessary: cross-training of personnel for consistent scoring of IHC reactivity, spot scoring of a fraction of the slides by a second individual, and calibration of equipment if different microscopes are being used. If care is taken to all steps throughout the process, IHC results can be robust and reproducible.

Assay Development of Multi-fluorophore Labeling Techniques

Multi-labeling methods provide valuable information for understanding the relationship between target proteins and distribution patterns. We developed an IHC-based multi-fluorophore labeling technique and used the technique in our tumor vascular and CTC studies. Tumor vasculature comprises endothelial cells, pericytes, basement membrane, and smooth muscles that are adjacent to tumor cells within a complicated and heterogeneous microenvironment. Multi-fluorophore labeling techniques allowed us to detect multiple target proteins using a selective antibody for each. We obtained data for the co-localization and expression of multiple biomarkers (target protein) and changes over time. Taken together, these data allowed us to understand how tumor vasculature was altered with drug treatment.

Tumor Vascular Measurement

During the tumor vascular study for linifanib development, we used IHC to identify biomarkers. We employed single antibody detection on serial sections to identify tissue structures, cell types, cellular components, and specific cell compartments, and used multi-fluorophore labeling techniques on serial sections to evaluate changes in tumor vasculature in response to the drug (Figs. 5.3 and 5.4). Multi-fluorophore labeling was used to assess microvessel density and diameter, integrity of the vascular wall, and hypoxic proteins within the same xenograft tumor tissue section, which provided assessment of tumor vasculature and helped answer critical questions for tumor response to the drug (Fig. 5.5). The multiple labeling technique included (1) lectin/fluorophore labeling of microvessels by perfusion of the vasculature (a non-Ag–Ab reaction in which lectin binds rapidly and uniformly to glycoprotein on luminal endothelial surfaces) [40–44], (2) a fluorophore-conjugated antibody specific to α-SMA to assess vascular maturation, or a fluorophore-conjugated antibody against collagen IV in the vessel basement membrane to determine integrity of the vascular wall and visualize vessel leakage, (3) detection of hypoxic protein expression by a fluorophore-conjugated monoclonal antibody against protein adducts derived by injection of pimonidazole, a substituted 2-nitroimidazole, which forms irreversible protein adducts at pO_2 levels ≤ 10 mmHg [45–48]. The Alexa dyes for conjugation to the primary Abs were selected to avoid the wavelength emission spectrum overlaps in the multi-fluorophore labeling procedure [25].

Vehicle   Linifanib

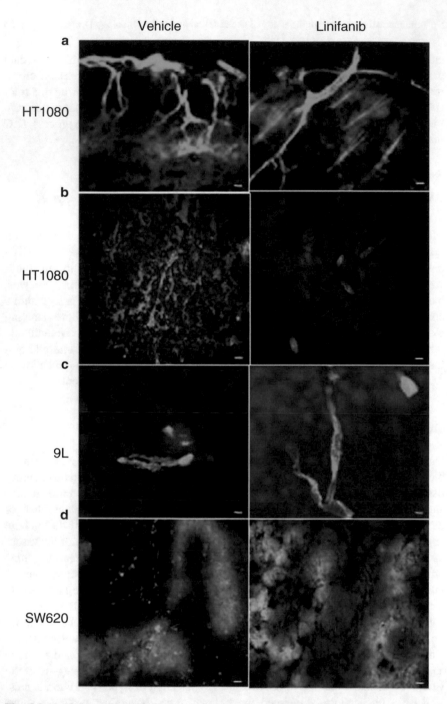

a

HT1080

b

HT1080

c

9L

d

SW620

Fig. 5.5 Multi-fluorophore labeling technique used in antiangiogenic drug development. (**a**) Lectin/FITC perfusion detected HT1080 global tumor vasculature. Linifanib treatment ablated chaotic microvessels. (**b**) Vessels stained with lectin/FITC (*green*) and pericytes stained with α-SMA (*red*) demonstrated that the drug improved pericyte coverage and reduced leakage. (**c**) Vessels stained with lectin/FITC (*green*) and basement membrane stained with collagen IV/Alexa 594 (*red*) indicated that the drug improved integrity of basement membrane and reduced leakage. (**d**) Hypoxic protein detected by mIgG/FITC (*green*) and microvessels stained with lectin/Alexa 594 indicated reduction of hypoxia and better organized vasculature with the treatment. *Bars* represent 100 μm in **a**, 20 μm in **b** and **d**, and 10 μm in **c**

To optimize the signal strength and minimize background from each assay, we planned the sequence and timing of the administration of pimonidazole to detect hypoxic protein expression, injection of lectin/fluorophore conjugate, and tumor collection. Briefly, tumor-bearing mice received an i.p. injection of pimonidazole hydrochloride 90 min before further hypoxia assessment, and subsequently the mice received an i.v. injection of lectin conjugated with a fluorophore. The dye was allowed to circulate for 5–10 min prior to tumor collection. All tumors were removed and snap frozen in liquid nitrogen and maintained at −80 °C until used. The cryosections were fixed in 4 % paraformaldehyde for 1 h, PBS washed, air dried, and stored at 4 °C for further IHC assay. 90 μm cryosections were cut for the observation of global vasculature (Fig. 5.5a). 20 μm cryosections were cut and further stained with either α-SMA/Alexa 594 or collagen IV/Alexa 594 for the observation of integrity of vascular wall (Fig. 5.5b, c), and the cryosections were stained with mouse IgG/APC for hypoxic protein expression (Fig. 5.5d). Using this technique, lectin, collagen IV, α-SMA, and hypoxic protein expression patterns were used to assess tumor vessels, the presence of vessel leakage, integrity of vascular walls, and tissue hypoxia level. Change in these vascular parameters correlated with tumor response to therapy.

Circulating Tumor Cell Assessment

We used multi-fluorophore labeling of specific antibodies to detect and enumerate CTC from cytological smears of cancer patient blood samples. ICC-based multi-fluorophore labeling allowed us to identify CTC from different tumor types and to develop stratification markers for patient selection while excluding unwanted blood cells. Using this technology in our poly(ADP-ribose)-polymerase (PARP) inhibitor ABT-888 development, we detected and enumerated CTC using a melanoma antibody cocktail labeled with an Alexa fluorophore followed by a second antibody to detect the stratification marker ERCC1 (excision repair cross-complementing 1) after a negative selection (excluded CD45+ leukocytes) during the process of CTC enrichment. If leukocytes are not eliminated during the CTC enrichment, a third antibody to detect CD45 should be used to help identification of CTC. The result provided valuable information for understanding the correlation of CTC numbers and changes of ERCC1 expression in CTC with melanoma patient response to treatment (data not shown [49]).

IHC Score

Scoring IHC assays is not a straightforward task, as various patterns may be observed in different cell and tissue types. The staining pattern, intensity, and area need to be carefully observed and incorporated into a scoring scheme. The current standard method for scoring protein expression by IHC is that stained features of cells, such as cytoplasmic, nuclear, or membrane staining, are classified as 0 (no staining), 1+ (weak staining), 2+ (moderate staining), and 3+ (strong staining) based on their intensity and completeness, and percentages of each pattern are integrated into the standard scoring scheme or the H score (a product of staining intensity and the percent of cells stained) [5, 39]. However, automated scanning microscopic systems detect the expression of biomarker proteins by measuring the intensity of Ab-conjugated fluorophores within a specified subcellular compartment in the tumor region and the result is a quantitative score of immunofluorescence intensity for the tumor. Automated detection provides more continuous and reproducible scoring of protein expression in tissue samples compared to manual evaluation [39, 50]. Nevertheless, trained professionals are still needed for the final judgment call in image analysis and morphological assessment.

Score for Tumor Vasculature

For the tumor vascular study using IHC in linifanib development, the staining patterns, including the cell types of the stain, the subcellular localization of the stain, relative intensity of the stain, and the percentage of cells of the stain were observed and captured by a scoring scheme (AxioVision 4.6, Carl Zeiss, Thornwood, NY) for evaluation.

Quantification of Microvessels and Target Receptors

Microvessel quantification, including density and diameter, was determined by the percentage of lectin/FITC perfused vessels as a range from 0 to 100 in area of $0.4 \, mm^2$. Quantification of the target receptor expressions (pPDGFR β and pVEGFR 2) was measured by the traditional pathology scoring system of 0–3: 0, no staining; (1) weak staining; (2) medium staining; and (3) strong staining (Fig. 5.4a and [26]).

Quantification of Hypoxic Protein Expression

Fluorescence intensities of specific hypoxic proteins were quantified by measuring the intensity staining over the background fluorescence (staining with only the fluorophore-conjugated secondary antibody). A fluorescence threshold was set to eliminate tissue autofluorescence. The fluorescence intensity of hypoxic protein represented the average brightness of all cell-related pixels (Fig. 5.5d and [25]).

Quantification of Pericyte Coverage

To assess vascular maturity, analysis of pericyte coverage on individual tumor vessels was performed by measuring the percentage of α-SMA+ cells surrounding lectin-stained vessels (Fig. 5.5b). The pericyte covered vessels were divided into three groups of 0–30 %, 50–60 %, and 80–100 % pericyte coverage. A pericyte coverage index was obtained by calculation of the average pericyte coverage per section, using the following equation (= number of 0 % coverage vessels/total number of vessels in a section × 0 % + number of 50 % coverage vessels/total number of vessels in a section × 50 % + number of 80 % coverage vessels/total number of vessels in a section × 80 %). These section-based values were further averaged for each tumor and compared between vehicle and ABT-869 treated groups [25].

Score for Circulating Tumor Cells

An automated cellular image system was used to scan the slides and detect rare events (Bioview system™, Israel). Enumeration and analysis of probably CTC is based on the presence of multi-fluorophore immunostaining, which indicates specific tumor marker binding, and the presence of appropriate morphological characteristics (Fig. 5.1c), such as large sized nuclei and increased ratio of nuclear and cytoplasm [38]. Cellular debris and cells with typical morphological features of normal hematological mononuclear cells are excluded from the analysis. Final CTC determination was performed by trained reviewers.

Statistical Analysis

In general, statistical analyses should be designed with statisticians prior to data collection after determining the characteristics and variables of the analyte to ensure that the appropriate number of tumor/slide/field is prepared and assessed to answer a specific question, as was performed for our tumor vascular study. For the final analysis, a mixed-effect model was fitted to estimate the changes of microvessel diameter and density for each group. Microvessel density values (number of vessels in the area of 0.4 mm^2) were log-transformed to achieve normality. The two-sample t-test was performed for the assessment of the average pericyte coverage per section to identify differences between the treatment groups, as well as vessel leakage and hypoxic protein quantification. For IHC staining intensity assessment, a Fisher Exact Test was used to compare the number of tumors with the scale ≥2 or <2. Values were expressed as Mean ± SE. Statistical analysis was carried out using SAS version 9.1 software. A p value of <0.05 was considered statistically significant.

The statistical analysis of CTC data was focused on CTC enumeration and the ratio of stratification marker expression level over the general CTC numbers. Clinical outcome data were combined when the data was analyzed to evaluate tumor response to the drug.

Companion Diagnostics

In oncology drug development, IHC assays can serve as a platform for developing companion diagnostics [2, 51]. IHC provides information on the localization of proteins and phospho-proteins in tissue sample. This can be particularly important for proteins with low or no abundance in serum. Assays such as HER-2/neu IHC and EGFR IHC are now routinely used to select treatments for breast and lung cancer, respectively [5, 7, 52, 53]. IHC is a standard working test in clinical labs and hospitals, and therefore new IHC assays can be readily implemented in a clinical setting.

Sources of IHC Errors and Troubleshooting

The IHC procedure and interpretation of the results can become complicated as more stringent goals of sensitivity and specificity are established. At least two factors need to be considered when IHC assays are performed: false negatives can occur if the antibody is inappropriately denatured, used at the wrong concentration, or if antigenicity in the sample is lost through poor fixation; false positives can occur if the antibody has nonspecific binding to the tissue, or presence of undissolved precipitates of chromagen, which can be corrected by filtering.

To prevent these errors, IHC methods should be optimized first through uniformity of tissue sample collection, fixation, paraffin removal (FFPE tissue), antigen retrieval (temperature, pH or protein digestion), minimizing background binding of HRP or ALP, proper selection of antibody (species and cross-activity), and determination of the working dilution. Secondly, IHC image capture for quantification should be consistent and representational across samples. A preliminary evaluation of image analysis tools for varying stain intensities on slides will help assure that analysis values are established optimally for all slides in the study.

Conclusions

IHC provides a useful approach for the development of cancer biomarkers. Classical immunostaining using chromagenic or fluorophore detection permits the identification and localization of proteins or other molecules in tissue sections and cytological preparation slides. However multi-fluorophore labeling techniques or a combination of antigen–antibody detection with non-Ag–Ab recognition should be given consideration for tumor vascular biomarker studies in antiangiogenic drug development and circulating tumor cell studies.

In this chapter, we described a general IHC method in detail and technical information on specific experiments to support oncology drug development, including as a tumor vascular biomarker study for linifanib, a multitargeted receptor tyrosine kinase inhibitor. Measurement of tumor vascular biomarkers by IHC may predict clinical outcome or response to anti-angiogenic therapy. Cancer angiogenic biomarkers include changes in target protein expression, microvessel density and diameter, integrity of the vascular wall, and tumor hypoxic protein expression. In preclinical tumor models, the anti-angiogenic effects measured by IHC correlated to inhibition of tumor growth. Linifanib had an immediate and significant effect on morphological and functional aspects of tumor vasculature, including decreases in tumor vascular permeability and microvessel density and diameter and increases in tumor vascular wall integrity [25, 26]. Additional analytical validation and qualification are still required for clinical application of the biomarkers.

The chapter also discussed using multi-fluorophore labeling ICC assay to detect and enumerate CTC and to assess the subcellular localization and expression pattern of proteins. Based on the results, the biomarkers may be useful for selecting an appropriate therapy for a particular cancer patient.

IHC assays are indispensable tools in the classification of tumors and in biomarker development. However, IHC methods must be developed with care and implemented by trained personnel to obtain robust results for use of the results in drug development. Factors which may contribute to uneven IHC results include specimen storage, duration of fixation, type of fixative, and conditions of tissue processing. Each element should be standardized for each IHC assay [9, 54, 55]. Finally, IHC assays must be rigorously designed with proper selection of the controls for the key variables, and they must be kept consistent when the procedure is performed in order to generate consistent, robust, and reproducible results.

Acknowledgement The authors thank Dr. Neela X. Patel for thoughtful suggestions and editing on this chapter.

References

1. Elias JM. Immunohistochemical methods. In: Elias JM, editor. Immunohistopathology. A practical approach to diagnosis. 2nd ed. Chicago, IL: ASCP; 2003.
2. Lynch F, Bernstein S. Immunohistochemistry assays in drug development performed by a contract research laboratory. In: Platero JS, editor. Molecular pathology in drug discovery and development. Hoboken, NJ: Wiley; 2009.
3. Ramos-Vara JA. Technical aspects of immunohistochemistry. Vet Pathol. 2005;42:405–26.
4. Wilson C, Schulz S, Waldman SA. Biomarker development, commercialization, and regulation: individualization of medicine lost in translation. Clin Pharm Ther. 2007;2:153–5.
5. Bartlett JMS, Going JJ, Mallon EA, Watters AD, Reeves JR, Stanton P, Richmond J, Donald B, Ferrier R, Cooke TG. Evaluating HER2 amplification and overexpression in breast cancer. J Pathol. 2001;195:422–8.
6. Pantel K, Brakenhoff RH, Brandt B. Detection, clinical relevance and specific biological properties of disseminating tumour cells. Nat Rev Can. 2008. doi:10.1038/nrc2375.
7. Santos GdC, Shepherd FA, Tsao MS. EGFR mutation and lung cancer: mechanisms of disease. Annu Rev Pathol. 2011;6:49–69.
8. Samoszuk MK, Walter J, Mechetner E. Improved immunohistochemical method for detecting hypoxia gradients in mouse tissues and tumors. J Histochem Cytochem. 2004;52(6):837–9.

9. Jain RK, Duda DG, Willett CG, Sahani DV, Zhu AX, Loeffler JS, Batchelor T, Sorensen G. Biomarkers of response and resistance to antiangiogenic therapy. Nat Rev Clin Oncol. 2009;6:327–38.
10. Li JL, Harris AL. The potential of new tumor endothelium-specific markers for the development of antivascular therapy. Cancer Cell. 2007;11:478–81.
11. Seaman S, Stevens J, Yang MY, Logsdon D, Graff-Cherry C, St. Croix B. Genes that distinguish physiological and pathological angiogenesis. Cancer Cell. 2007;11:539–54.
12. Sharma RA, Harris AL, Galgleish AG, Steward WP, O'Byrne KJ. Angiogenesis as a biomarker and target in cancer chemoprevention. Lancet Oncol. 2001;2:726–32.
13. Faccioli N, Marzola P, Boschi F, Sbarbati A, D'Onofrio M, Mucelli RP. Pathological animal models in the experimental evaluation of tumour microvasculature with magnetic resonance imaging. Radiol Med. 2007;112:319–28.
14. Sikora J, Dworacki G, Trybus M, Batura-Gabryel H, Zeromski J. Correlation between DNA content, expression of Ki-67 antigen of tumor cells and immunophenotype of lymphocytes from malignant pleural effusions. Tumor Biol. 1998;19:196–204.
15. Trusheim MR, Berndt ER, Douglas FL. Stratified medicine: strategic and economic implications of combining drugs and clinical biomarkers. Nat Rev. 2007;6:287–93.
16. Batchelor TT, Sorensen AG, Tomaso ED, Zhang WT, Duda DG, Cohen KS, Kozak KR, Cahill DP, Chen PJ, Zhu MW, Ancukiewicz M, Mrugala MM, Plotkin S, Drappatz J, Louis DN, Ivy P, Scadden DT, Benner T, Loeffler JS, Wen PY, Jain RK. AZD2171, a Pan-VEGF receptor tyrosine kinas inhibitor, normalizes tumor vasculature and alleviates edema in glioblastoma patients. Cancer Cell. 2007;11:83–95.
17. Brustmann H, Riss P, Naude S. The relevance of angiogenesis in benign and malignant epithelial tumor s of the ovary: a quantitative histological study. Gynecol Oncol. 1997;67:20–6.
18. Chiba Y, Taniquchi T, Matsuyama K, Sasaki M, Kato Y, Tanaka H, Muraoka R, Tanigawa N. Tumor angiogenesis, apoptosis and p53 oncogene in stage I lung adenocarcinoma. Surg Today. 1999;29:1148–53.
19. Fox SB, Leek RD, Weekes MP, Whitehouse RM, Gatter KC, Harris AL. Quantitation and prognostic value of breast cancer angiogenesis: comparison of microvessel density, Chalkley count, and computer image analysis. J Pathol. 1995;177:275–83.
20. Weidner N, Carroll PR, Flax J, Blumenfeld W, Folkman J. Tumor angiogenesis correlates with metastasis in invasive prostate carcinoma. Am J Pathol. 1993;143:401–9.
21. Kononen J, Bubendorf L, Kallioniemi A, Barlund M, Schraml P, Leighton S, Torhorst J, Mihatsch MJ, Sauter G, Kallioniem OP. Tissue microarrays for high-throughput molecular profiling of tumor specimens. Nat Med. 1998;4:844–7.
22. Wan WH, Fortuna MB, Furmanski P. A rapid and efficient method for testing immunohistochemical reactivity of monoclonal antibodies against multiple tissue samples simultaneously. J Immunol Methods. 1987;103:121–9.
23. Hayat MA. Fixation and embedding. In: Hayat MA, editor. Microscopy, immunohistochemistry, and antigen retrieval methods for light and electron microscopy. New York: Kluwer Academic; 2002.
24. Dapson RW. Fixation for the 1990's: a review of needs and accomplishments. Biotech Histochem. 1993;68:75–82.
25. Jiang F, Albert DH, Luo YP, Tapang P, Zhang K, Davidsen SK, Fox GB, Lesniewski R, McKeegan EM. ABT-869, a multi-targeted receptor tyrosine kinase inhibitor, reduces tumor microvascularity and improves vascular wall integrity in preclinical tumor models. JPET. 2011;338:134–42.
26. Luo Y, Jiang F, Cole TB, Hradil VP, Reuter D, Chakravartty A, Albert DH, Davidsen SK, Cox BF, McKeegan EM, Fox GB. A novel multi-targeted tyrosine kinase inhibitor, linifanib (ABT-869), produces functional and structural changes in tumor vasculature in an orthotopic rat glioma model. Cancer Chemother Pharmacol. 2012;69(4):911–21 [Epub ahead of print], PMID:22080168 [PubMed—as supplied by publisher].
27. Eltoum I, Fredenburgh J, Myers RB, Grizzle WE. Introduction to the theory and practice of fixation of tissues. J Histotechnol. 2001;24:173–90.

28. Shi SR, Cote RJ, Taylor CR. Antigen retrieval techniques: current perspectives. J Histochem Cytochem. 2001;49:931–8.
29. Riera JR, Atengo-Osuna C, Longmate JA, Battifora H. The immunohistochemical diagnostic panel for epithelial mesothelioma: a reevaluation after heat-induced epitope retrieval. Am J Surg Pathol. 1997;12:1395–8.
30. Burry RW. Specificity controls for immunocytochemical methods. J Histochem Cytochem. 2000;48(2):163–5.
31. Albert DH, Tapang P, Magoc TJ, Pease LJ, Reuter DR, Wei RQ, Li J, Guo J, Bousquet PF, Ghoreishi-Haack NS, Wang B, Bukofzer GT, Wang YC, Stavropoulos JA, Hartandi K, Niquette AL, Soni N, Johnson EF, McCall JO, Bouska JJ, Luo Y, Donawho CK, Dai Y, Marcotte PA, Glaser KB, Michaelides MR, Davidsen SK. Preclinical activity of ABT-869, a multi targeted receptor tyrosine kinase inhibitor. Mol Cancer Ther. 2006;5(4):995–1006.
32. Taylor CR, Shi S-R, Barr NJ, Wu N. Techniques of immunohistochemistry: principles, pitfalls, and standardization. In: Dabbs DJ, editor. Diagnostic immunohistochemistry. New York: Churchill Livingstone; 2002.
33. Hsu SM, Raine L, Fanger H. Use of avidin-biotin-peroxidase complex (ABC) in immunoperoxidase techniques: a comparison between ABC and unlabeled antibody (PAP) procedures. J Histochem Cytochem. 1981;29:577–80.
34. Xu J, Stolk JA, Zhang X, Silva SJ, Houghton RL, Matsumura M, Vedvick TS, Leslie KB, Badaro R, Reed SG. Identification of differentially expressed genes in human prostate cancer using subtraction and microarray. Cancer Res. 2000;60:1677–82.
35. Cristofanilli M, Budd GT, Ellis MJ, Stopeck A, Matera J, Miller MC, Reuben JM, Doyle GV, Allard WJ, Terstappen LWMM, Hayes DF. Circulating tumor cells, disease progression, and survival in metastatic breast cancer. N Engl J Med. 2004;351:781–91.
36. Giordano A, Giuliano M, De Laurentiis M, Arpino G, Jackson S, Handy BC, Ueno NT, Andreopoulou E, Alvarez RH, Valero V, De Placido S, Hortobagyi GN, Reuben JM, Cristofanilli M. Circulating tumor cells in immunohistochemical subtypes of metastatic breast cancer: lack of prediction in HER2-positive disease treated with targeted therapy. Ann Oncol. 2012;23(5):1144–50 [Epub ahead of print].
37. Gandhi L, Camidge DR, Ribeiro de Oliveira M, Bonomi P, Gandara D, Khaira D, Hann CL, McKeegan EM, Litvinovich E, Hemken PM, Dive C, Enschede SH, Nolan C, Chiu YL, Busman T, Xiong H, Krivoshik AP, Humerickhouse R, Shapiro GI, Rudin CM. A phase 1 study of Navitoclax (ABT-263), a novel Bcl-2 family inhibitor, in subjects with small cell lung cancer (SCLC) and other solid tumors. J Clin Oncol. 2011;29(7):909–16.
38. Bauer KD, Torre-Bueno JDL, Diel IJ, Hawes D, Decker WJ, Priddy C, Bossy B, Ludmann S, Yamamote K, Masih AS, Espinoza FP, Harrington DS. Reliable and sensitive analysis of occult bone marrow metastases using automated cellular imaging. Clin Can Res. 2000;3552(6):3552–9.
39. McCabe A, Dolled-Filhart D, Camp RL, Rimm DL. Automated quantitative analysis (AQUA) of in situ protein expression, antibody concentration, and prognosis. J Natl Cancer Inst. 2005;97(24):1808–15.
40. Debbage PL, Griebel J, Ried M, Gneiting T, DeVries A, Hutzler P. Lectin intravital perfusion studies in tumor-bearing mice: micrometer-resolution, wide area mapping of microvascular labeling, distinguishing efficiently and inefficiently perfused microregions in the tumor. J Histochem Cytochem. 1998;46:627–39.
41. Gee MS, Procopio WN, Makonnen S, Feldman MD, Yeilding NM, Lee WMF. Tumor vessel development and maturation impose limits on the effectiveness of anti-vascular therapy. Am J Pathol. 2003;162:183–93.
42. Hashizume H, Baluk P, Morikawa S, McLean JW, Thurston G, Roberge S, Jain RK, McDonald DM. Openings between detective endothelial cells explain tumor vessel leakiness. Am J Pathol. 2000;156:363–80.
43. Minamikawa T, Miyake T, Takamatsu T, Fujita S. A new method of lectin histochemistry for the study of brain angiogenesis. Histochemistry. 1987;87:317–20.

44. Morikawa S, Baluk P, Kaidoh T, Haskell A, Jain RK, McDonald DM. Abnormalities in pericytes on blood vessels and endothelial sprouts in tumors. Am J Pathol. 2002;160:985–1000.
45. Dings RPM, Loren M, Heun H, McNeil E, Griffioen AW, Mayo KH, Griffin RJ. Scheduling of radiation with angiogenesis inhibitors anginex and avastin improves therapeutic outcome via vessel normalization. Clin Cancer Res. 2007;13:3395–402.
46. Raleigh JA, Chou SC, Arteel GE, Horsman MR. Comparisons among pimonidazole binding, oxygen electrode measurements, and radiation response in C3H mouse tumors. Radiat Res. 1999;151:580–9.
47. Raleigh JA, Chou SC, Calkins-Adams DP, Ballenger CA, Novotny DB, Varia MA. A clinical study of hypoxia and metallothionein protein expression in squamous cell carcinomas. Clin Cancer Res. 2000;6:855–62.
48. Varia MA, Calkins-Adams DP, Rinker LH, Kennedy AS, Novotny DB, Fowler Jr WC, Raleigh JA. Pimonidazole: a novel hypoxia marker for complementary study of tumor hypoxia and cell proliferation in cervical carcinoma. Gynecol Oncol. 1998;71:270–7.
49. Middleton M, Friedlander P, Hamid O, Daud A, Plummer R, Falotico N, Chyla B, Jiang F, McKeegan E, Mostafa NM, Zhu M, Qian J, McKee M, Luo Y, Giranda VL, McArthur GA. Randomized phase II study evaluating veliparib (ABT-888) with temozolomide in patients with metastatic melanoma. Ann Oncol. 2015;26:2173–9.
50. Camp RL, Chung GG, Rimm DL. Automated subcellular localization and quantification of protein expression in tissue microarrays. Nat Med. 2002;8:1323–7.
51. Food and Drug Administration. Medical device: guidance for submission of immunohistochemistry applications to the FDA; final guidance for industry. 2008. http://www.fda.gov/MedicalDevices/DeviceRegulationandGuidance/GuidanceDocuments/ucm094002.htm
52. Mass R. The role of HER-2 expression in predicting response to therapy in breast cancer. Semin Oncol. 2000;27:46–52.
53. Ravdin PM. Should HER2 status be routinely measured for all breast cancer patients? Semin Oncol. 1999;26:117–23.
54. Bertheau P, Cazals-Hatem D, Meignnin V, de Roquancourt A, Verola O, Lesourd A, Sene C, Brocheriou C, Janin A. Variability of immunohistochemical reactivity on stored paraffin slides. J Clin Pathol. 1998;51:370–4.
55. Jacobs TW, Prioleau JE, Stillman IE, Schnitt SJ. Loss of tumor marker-immunostaining intensity on stored paraffin slides of breast cancer. J Natl Cancer Inst. 1996;88:1054–9.

Chapter 6
Male Reproductive Toxicology and the Role of Immunohistochemistry

Daniel G. Cyr

Introduction

Several reports have suggested deterioration in the quality of male human reproductive function over the last 50 years [1–5]. Male reproductive parameters that have been reported as altered include a decrease in sperm count and increased incidence of congenital malformations of the male reproductive tract and of testicular cancer among young males [5]. While it is not clear what factors or changes in lifestyle may be responsible for these changes, it has been suggested that exposure to certain environmental contaminants may be contributing to the rising rates of these [5]. Of the different classes of reproductive toxicants present in the environment, those that act as endocrine-disrupting chemicals have been singled out as contributing to male reproductive dysfunction [5]. A wide variety of industrial, pharmaceutical, agricultural, and natural compounds have been implicated in reproductive toxicology [6–8]. However, not all reproductive toxicants act via the disruption of the endocrine system. Several studies have also shown that environmental toxicants can alter epididymal function and development, which may, in turn, have important consequences on sperm quality [8, 9].

The goal of male reproduction is the formation of healthy spermatozoa. The production of spermatozoa by spermatogenesis in the testis is complex, and requires key cellular interactions between developing germ cells and between germ cells and Sertoli cells [10, 11]. Likewise, sperm maturation in the epididymis depends upon the formation of the blood–epididymis barrier, which permits the formation of a

D.G. Cyr (✉)
INRS-Institut Armand-Frappier, University of Quebec, 531 Boulevard des Prairies,
Laval, QC, Canada, H7B 1B7
e-mail: Daniel.Cyr@iaf.inrs.ca

© Springer Science+Business Media New York 2016
S.A. Aziz, R. Mehta (eds.), *Technical Aspects of Toxicological Immunohistochemistry*, DOI 10.1007/978-1-4939-1516-3_6

95

Fig. 6.1 Schematic
representation of the testis
and epididymis

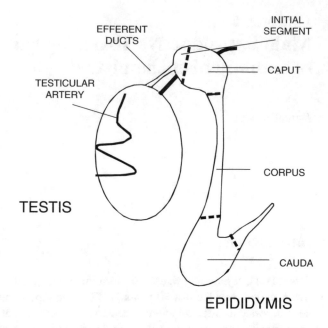

specific luminal environment necessary for sperm maturation (Fig. 6.1) [12]. In addition, one of the major roles of the epididymis is to protect spermatozoa from toxicants, since they have limited capacity to defend themselves from chemical insult. Understanding the effects of toxicants on the male necessitates specific markers which provide not only tissue-specific indices of effects but also cell-specific tools to understand the targets and mechanisms by which toxicants alter either testicular functions or epididymal sperm maturation.

The Testis

The primary function of the testis is to form spermatozoa and synthesize steroid hormones. These functions are accomplished by the seminiferous tubules, which are responsible for the production of spermatozoa, and the interstitial Leydig cells which synthesize steroids [10]. The seminiferous tubules are enveloped by a limiting membrane composed of peritubular myoid cells as well as monocytes [10]. The epithelium of the seminiferous tubules is composed of Sertoli cells and developing germ cells, which are formed by spermatogenesis and spermiation. At the end of spermiation, spermatozoa are released into the lumen of the tubules, coiled structures which fuse into the rete testis [13].

Testicular functions are regulated primarily by the hypophyseal hormones: luteinizing hormone (LH), which regulates steroidogenesis by Leydig cells, and follicle-stimulating hormone (FSH), which is implicated in Sertoli cell development.

In addition, thyroid hormones (TH) are critical for Sertoli cell differentiation. In the absence of TH, Sertoli cells maintain an immature appearance and continue to proliferate [14–19]. TH act on Sertoli cells to induce the expression of the gap junction protein Cx43, which induces cellular differentiation [20–23].

Spermatogenesis

The process by which spermatogonia develop and differentiate into spermatozoa is referred to as spermatogenesis. This process involves a complex interplay between the developing germ cells and between germ cells and Sertoli cells. The timing of spermatogenesis differs among species and can range from approximately 65 days in the rat to 85 days in humans [10]. This is a well-defined process characterized by a proliferative phase, in which spermatogonia divide by mitosis, a meiotic phase in which spermatocytes undergo two meiotic divisions, leading to the formation of haploid spermatids, and a morphological differentiation phase, or spermiogenesis, in which the cells develop a flagella, an acrosome, and a compacted nucleus, as a result of the replacement of histones by protamines, which bind the chromatin [10, 24].

Histological sections of the seminiferous tubule indicate that developing germ cells are arranged in well-defined associations. These associations can be grouped into different stages of development, which are referred to as the stages of the cycle of spermatogenesis. In the rat, 14 different stages of spermatogenesis have been distinguished and these can be used to assess perturbations in the process of spermatogenesis [25]. The stages of the cycle occur sequentially along the length of the seminiferous tubule. Since there are 4–5 generations of germ cells at each stage of development, each stage reflects the maturation of a particular generation of germ cells. For example, spermatozoa are released into lumen of the testis at stage VIII of spermatogenesis [25]. This requires specific proteins and a spatiotemporal expression so that these may exert their specific roles during spermatogenesis. Thus, the use of immunolocalization of these proteins with respect to the stages of spermatogenesis represents critical knowledge to assess the effects of toxicants on the testis of adults.

Spermatogenesis begins during postnatal development. In the rat testis, pachytene spermatocytes are first observed after the formation of the blood–testis barrier, which occurs between 14 and 15 days of age [26]. The first wave of spermatogenesis is completed at approximately day 42. This coincides with increasing blood testosterone levels, which peak at approximately day 39 [27].

As is the case in most tissues with cell proliferation, spermatogenesis is a delicate balance between cell proliferation and apoptosis. Gestational exposure to dibutyl phthalates was reported to cause a decrease in the number of germ cells in rat fetuses late in gestation by inhibiting cell proliferation [28, 29]. The effect, however, was transient, and germ cell number in exposed rats was similar to vehicle-exposed controls 2 days after birth [28, 29]. In contrast, exposure of rats to mono-(2-ethylhexyl) phthalate (MEHP) induced germ cell apoptosis by increasing the expression of FasL in Sertoli cells. This increase was shown to be mediated via the NF-kappaB pathway which, along with the Sp-1 transcription factor, bound to specific response elements on the FasL promoter to increase its expression [30]. Experiments examining the effects of toxicants that target the Sertoli cells, such as MEHP, 2,5 hexanedione (2,5HD), and carbendazim, were all shown to induce germ cell apoptosis in the testis, as determined by Tunel staining [31]. However, the mechanisms of toxicity and the parameters that resulted from exposure to any of these toxicants were varied, suggesting that standardized histopathological assessment of the testis may be difficult to achieve.

Assessing effects on the development of the testis may also be complicated and the timing of exposure may be important in identifying effects on specific windows of testicular development. In utero exposure to tributyltin (TBT) from gestational days 0–19 or 8–19 resulted in a decreased number of Sertoli cells at the highest dose (20 mg/kg bw/day) and for both exposure time periods, but gonocyte populations decreased only in rats treated from gestational days 8–19 [32]. In the intertubular region between adjacent interstitial cells, immunostaining for the gap junctional protein Cx43 was reduced or completely absent in treated rats. This suggests the need of more than one histopathological approach to understand the effects of toxicants on the developing testis [32].

The Blood–Testis Barrier

The differentiation of Sertoli cells culminates with the formation of the blood–testis barrier [33]. The barrier is composed of basally located tight junctions between neighboring Sertoli cells of the seminiferous tubules. Tight junctions are closely associated plasma membranes from adjacent cells that form an impermeable barrier to large molecules. The tight junctions between Sertoli cells comprise occludin and claudins (Cldns), which are transmembrane proteins. While initial studies identified several claudins in the testis, Gow et al. [34] showed, using transgenic mice, that the blood–testis barrier was dependent on the presence of Cldn11. In the absence of Cldn11, the barrier was compromised and spermatogenesis was arrested. In androgen receptor-null mice, Meng et al. [35] reported that the blood–testis barrier was also compromised and there was a decrease in Cldn3 levels. Whether or not the decreased levels of Cldn3 were the cause, or a consequence, of loss of barrier function in the testis remains to be established. In occludin-null mice, the first waves of spermatogenesis appeared to be normal, although over time the males became infertile. Clearly, alterations in the expression and localization of either occludin or specific Cldns appear to represent critical markers for Sertoli cell barrier function, since these proteins can be sensitive targets of toxicants [36].

The formation of testicular tight junctions and the blood–testis barrier are complex and involve both cell adhesion molecules, in particular neural cadherin (cdh2), and connexin43 (Cx43)-mediated gap junctions. Cadherins are single-pass transmembrane proteins that mediate calcium-dependant cell adhesion [12, 37, 38].

They are responsible for the formation of the zonula adherens around the cells of epithelia. Cyr et al. [39] first described the presence of Cdh2 and placental cadherin (Cdh3) in the rat testis. In the mouse, Cdh3 is expressed in gonocytes and weakly expressed in Sertoli cells during development [40]. In the adult, Cdh3 is localized to the peritubular region of the seminiferous tubule and Sertoli cells [41], and it has been reported in spermatogonia and spermatocytes [42]. In contrast to Cdh3, Cdh2 levels increase during the first wave of spermatogenesis and Cdh2 is localized to the surface of spermatogonia and primary spermatocytes, and possibly early spermatids [39, 43]. Johnson and Boekelheide [44] reported that Cadh2 was present between spermatocytes and Sertoli cells in the rat. In vitro studies showed that Cd disrupted tight junctions between Sertoli cells. This effect was accompanied by decreased levels of occludin and cellular relocalization of E-cadherin [45, 46].

Gap junctions mediate intercommunication between neighboring cells and are critical for cell proliferation and differentiation. They are composed of connexons, or hemichannels, contributed by each of the neighboring cells. Connexons comprise six connexins, and the contribution of Cxs confers selectivity of the pores between the cells [38, 47]. Gap junctions are responsible for intercellular communication by forming intercellular pores between cells that allow for the movement of small molecules (<1 kDa), including secondary messengers (e.g., cAMP), between cells. Gap junctions have been shown to be highly sensitive to chemical insult in many tissues, particularly the liver [48, 49]. In the testis, Cx43 colocalizes with occludin in Sertoli cells in the area of the blood–testis barrier [50]. Several studies have shown that Cx43 is important for the formation tight junctions between Sertoli cells [51–54]. Li et al. [53] have reported that Cx43, along with plakophilin2, can alter the transepithelial resistance of Sertoli cells in vitro, suggesting that these two proteins interact in the regulation of tight junctions. However, in mice lacking Sertoli cell-specific Cx43, a blood–testis barrier was formed and shown to be functional [51]. Furthermore, levels of occludin, N-cadherin, and β-catenin were all increased, while ZO-1 levels were decreased. Similar results were obtained in Sertoli cells in which Cx43 levels were knocked down using antisense siRNA. While these results suggest that having a functional blood–testis barrier is not the only requirement for spermatogenesis to proceed, the question arises as to whether or not the barrier is, in fact, fully patent and functional. Alterations in adherens junctions indicate that the dynamic nature of the blood–testis barrier is altered in Cx43 Sertoli-null mice. The tight junctions of the blood–testis barrier are unique in that developing germ cells must be transported across the tight junctions from the basal to the adluminal compartment. Thus, the dynamic nature of testicular tight junctions requires precise interactions between various types of cellular junctions to coordinate these processes. However, the question remains as to how Cx43 gap junctions accomplish this; perhaps the asymmetric intercellular communication between Sertoli cells and germ cells reported by Risley et al. [55] activates signaling from the germ cells back to the Sertoli cells to regulate both the adherens and tight junctions of the blood–testis barrier.

Studies by Aravindakshan et al. [56] showed that lactating female rats fed fish captured at a site where endocrine-disrupting chemicals induced estrogenic effects

Cx43

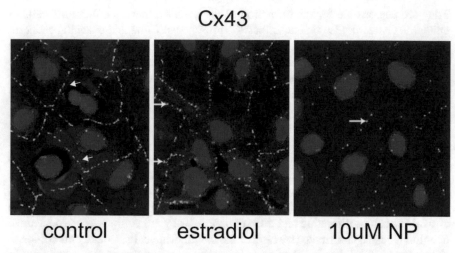

control   estradiol   10uM NP

Fig. 6.2 Photomicrographs of TM4 murine Sertoli cells showing the immunolocalization of Cx43. In control cells, punctate Cx43 immunofluorescence (*arrows*) is prominent at the plasma membranes between adjacent cells. Cells treated for 24 hrs with 17β-estradiol did not show signifi-cant changes in Cx43 staining. In cells incubated with 10 μM of nonylphenol for 24 hrs, Cx43 immunostaining (*green*) is dramatically decreased (*arrows*). Nuclei are stained with propidium iodide (*red*) (×600)

in fish resulted in alterations in spermatogenesis in the pups when they became adult. This effect was accompanied by decreased immunoreaction of Cx43 between Sertoli cells of the testis. To assess whether or not the effects on the testis were due to estrogenic chemicals, Aravindakshan and Cyr [57] exposed murine Sertoli cells (TM4 cells) to varying doses of nonylphenol, a prominent contaminant near the site where the fish were captured, and estradiol. Both intercellular gap junctional communication assays and Cx43 immunofluorescence studies indicated a dose-dependent response with increased concentrations of nonylphenol. However, estra-diol exposure had no effect on either Cx43 levels or intercellular communication, indicating that the effects on gap junctions mediated by nonylphenol did not occur via an estrogenic signaling pathway (Fig. 6.2).

The Epididymis

Spermatozoa from the testis are immotile and cannot fertilize. They must transit through the epididymis, where they acquire the ability to both swim and fertilize, a process referred to as sperm maturation. Testicular input to the epididymis is con-veyed via 3–5 efferent ductules which anastomose to form a unique and highly convoluted epididymal duct. Early studies by Bedford [58] and Orgebin-Crist [59] indicated that epididymal sperm maturation was due to the exposure and interaction of the spermatozoa with components of the fluid of the epididymal lumen. Thus, the ability of the epididymis to provide the appropriate milieu for sperm maturation is critical; this is dependent on the polarized functions of the epithelial cells that secrete and remove proteins and other biomolecules into the lumen. The luminal environment is also dependent on the presence of the blood–epididymis barrier, which is formed by tight junctions between epithelial cells that line the lumen of the epididymis [12, 60].

Structure of the Epididymis

In most species, the epididymis comprises four segments identified according to functional and morphological differences: the initial segment, the caput, the corpus, and the cauda epididymidis. In the human, there is no morphologically distinct initial segment as the proximal caput region appears to comprise primarily of efferent ducts [61]. Whether or not some of the physiological functions of the initial segment exist in the efferent ducts of the human or immediately adjacent to these in the human epididymis is not known.

In all species, the basic structure of the epididymal epithelium remains very similar and is composed of a pseudostratified columnar epithelium containing principal, clear, and basal cells throughout the epididymis. Narrow and apical cells are present exclusively in the initial segment region. Principal cells are the most abundant cell type of the epididymis and are involved in protein secretion and absorption. They have an extensive stereocilia and an elaborate Golgi network. Clear cells are dispersed between principal cells in the epithelium and their numbers increase along the length of the epididymis from the proximal to distal regions [62]. These cells have poorly developed stereocilia and are primarily involved in absorption of materials from the lumen. Basal cells are located at the base of the epithelium and have long thin projects that interact with other basal cells and may extend apically to the lumen. Basal cells are thought to have a role in immunity and/or in the regulation of principal cell functions [63–65].

The Blood–Epididymis Barrier

The ultrastructure of tight and gap junctions in the epididymis was first described by Friend and Gilula [66] in the rat. They reported the presence of an extensive tight junction between adjacent principal cells localized to the apical region of the cells. Pelletier [67] reported the presence of tight junctions between principal and clear cells in the mink epididymis. Using thin section electron microscopy, Cyr et al. [68] demonstrated that the length of the junctional complex in the rat is reduced along the lateral plasma membrane of principal cells in the corpus and cauda, as compared with the initial segment.

Fig. 6.3 Immunocytochemical localization of Cldn1 in the caput region of the epididymis. Cldn1 was localized to the apical region of the epithelium where tight junctions are located (*black arrows*). Immunostaining was also observed along the lateral margins of the plasma membrane (*yellow arrow*) and between basal and principal cells (*white arrow*). *P* principal cell, *B* basal cell, *Lu* lumen, *n* nucleus

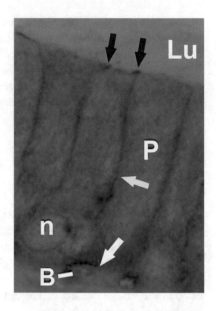

Using immunofluorescence and confocal microscopy, Cyr et al. [50] showed that the tight junction protein occludin was expressed as early as embryonic day 13.5 in the murine epididymis. At this age, however, occludin was localized primarily in the cytoplasm. By embryonic day 18.5, however, occludin was localized to the lateral margins of plasma membrane of epithelial cells. Occludin is localized to the apical cell surface of the mouse epididymis by embryonic day 18.5, which coincides with peak androgen levels, suggesting that androgens may be important in the regulation of embryonic epididymal tight junctions. The immunolocalization of occludin correlated well with studies of Suzuki and Nagano [69] who had reported that the number of tight junctional strands in the apical region of the epididymis increased as a function of age. Cldns are also present between adjacent principal in the area of the tight junction and, in the case of Cldn1, along the lateral margins of cell [70, 71] (Fig. 6.3). The importance of cell–cell interactions in the developing epithelium is critical to toxicology with respect to assessing the effects of toxicants on these interactions. Tight junctions have been shown to be sensitive targets for environmental toxicants, which affect primarily the levels of proteins implicated in the formation of tight junctions.

In addition to tight junctions, gap junctions are present in the apical region of the epithelium and were observed by freeze fracture electron microscopy to be intermixed with the tight junctions of the blood–epididymis barrier [66]. Pelletier [67], using freeze fracture electron microscopy in the mink, a seasonal breeder, showed that gap junctions were present basal to the tight junctions but also at the basal region of principal cells. He also reported that gap junctions were present between principal and clear cells. Cyr et al. [72] using immunohistochemistry showed that connexin 43 (Cx43) was localized between principal and basal cells of the adult epididymis. Furthermore, their results showed that the cellular localization of Cx43 varied as a function of circulating levels of androgens. In castrated or orchidecto-mized rats, Cx43 was immunolocalized to the apical region of the epididymal epithelium between adjacent principal cells, as well as between basal and principal cells. In castrated animals administered testosterone, Cx43 remained localized between principal and basal cells and no apical staining was observed. Thus, its localization has potential for identifying the effects of anti-androgenic endocrine-disrupting chemicals on epididymal function since the cellular localization of Cx43 varies as a function of androgens in the epididymis.

Dufresne et al. [73] identified other epididymal Cxs, Cx26, 32, 31.1, and 30.3, in addition to Cx43. Unlike Cx43, both Cx26 and Cx32 were localized between adja-cent principal cells, using immunofluorescent techniques in cryosections of the rat epididymis. These two Cxs were colocalized in young rats, but their localization along the lateral plasma membrane of epididymal epithelial cells was distinct in adult rats [73].

Markers of Epididymal Function

There have been relatively few toxicology studies that have focused on epididymal function from the perspective of assessing changes in immunolocalization of spe-cific proteins. However, since the epididymis is critical to sperm maturation, and requires well-coordinated functions along the epididymal tubule in order to regulate sperm maturation, the use of immunohistochemistry markers to assess the function of specific cell types within the epithelium of the epididymis is crucial to under-standing toxicant-induced alterations to cauda epididymal sperm.

The most abundant cell type of the epididymis is the principal cell. These cells line the lumen of the epididymis and are involved in the secretion and reabsorption of proteins from the lumen. Several proteins secreted by the principal cells can be used to examine the secretory capacity of principal cells and these have been previ-ously reviewed. A detailed list of secreted proteins, cell types, and which epididy-mal region secretes these proteins has been produced [74].

Basal cells are located at the base of the epididymal epithelium. Using immuno-fluorescence and confocal microscopy with three-dimensional reconstruction, it was possible to demonstrate that these cells have extensive lateral projections that form a basket-like structure [63, 64, 75]. Immunohistochemical studies indicated that basal cells express superoxide dismutase-1 (SOD1), glutathione-s-transferase (GST), and metallothioneins (Fig. 6.4) [63, 76, 77]. Studies with GST immunohis-tochemistry suggested that basal cells may have projections that extend apically. A recent confocal microscopy study using paraformaldehyde-fixed thick sections and antisera against the tight junction protein Cldn1 indicated that basal cell projections may reach the lumen of the epididymis. The authors suggested that these may act as epididymal sensors. Given the fact that these cells express a number of proteins and enzymes implicated in the scavenging of free radicals, it appears likely that one of

Fig. 6.4 Immunocytochemical
localization of metallothionein
in the rat epididymis. In the
cauda epididymidis of adult
rats, metallothionein was
immunolocalized in basal cells
(*arrowheads*). No reaction was
observed over principal cells
or spermatozoa in the
epididymal lumen. *P* principal
cells, *Lu* lumen, *IT* interstitium

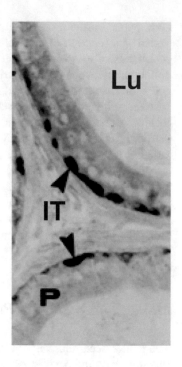

the functions of these cells is to protect the epithelium, and possibly maturing spermatozoa, from free radicals. While heavy metals, such as mercury, can induce
metallothionein in the epididymis, few studies have examined the role of basal cells
in the scavenging of free radicals following chemical insult. Clearly, however, these
cells may play an integral role in detoxification in the epididymis.

Clear cells are endocytic cells present in the caput corpus and cauda epididymidis. Compared to principal cells, the cells have poorly developed stereocilia and
contain a vast number of endosomes, multivesicular bodies, and lysosomes. The
cells endocytose the cytoplasmic droplets that are released from maturing spermatozoa during epididymal transit in the corpus epididymidis [62]. Clear cells are also
implicated in the regulation of luminal acidification via a number of proton-
secreting ATPases (V-ATPase) which are activated via bicarbonate-sensitive ade-
nylate cyclases [78].

Many of the physiological functions of the epididymis are highly dependent on
the actions of androgens. As with the ventral prostate and seminal vesicles, the
weight of the epididymis is highly correlated with circulating levels of androgens
[79, 80]. The level of testosterone in the luminal fluid of the epididymis is approximately 10 times higher than circulating levels [81]. The androgen-metabolizing
enzyme 5α-reductase, which converts testosterone to dihydrotestosterone, is present
in epididymal principal cells, particularly in the initial segment, which contains
high levels of 5α-reductase type 1while type 2 distributed all along the epididymis.
Androgen receptors are present in both principal and basal cells throughout the

epididymis [81]. Several genes have been identified that are either androgen stimulated or androgen repressed [79, 81]. Many of these have been identified immunohistochemically and represent markers that could be used to assess the effects of toxicants on androgen action. For example, clusterin (also known as sulfated glycoprotein-2) is an androgen-repressed gene [82]. It is expressed in the principal cells of the distal initial segment and caput and weakly expressed in the corpus and cauda epididymidis [83]. Orchidectomy and orchidectomy with androgen replacement experiments have indicated that both carbonic anhydrase (CA) II, which is expressed throughout the epididymis, and CAIV, which is expressed in the corpus, are androgen-dependent genes [84]. The cellular distribution of certain junctional proteins has been shown to vary in response to the absence of testosterone in orchidectomized rats. For example, Cx43, which is localized in normal intact animals between principal and basal cells, is localized not only between basal and principal cells in orchidectomized rats but also between adjacent principal cells. In rats whose testosterone levels have been maintained using subdermal implants, Cx43 is localized exclusively between basal and principal cells [72]. These observations highlight the importance of using an immunohistochemical approach to assess the effects of hormones or toxicants on epididymal proteins.

Epididymal Cell Lines

Several in vitro models have been used to study the epididymis. Most of the in vitro models initially relied on organ cultures, tubule cultures, or primary cell cultures. Telgmann et al. [85] developed one of the first immortalized epididymal cell lines from the dog by transfecting primary epididymal cultures with plasmids expressing either the SV40 Large-T antigen (SV40 LTAg) or c-*myc*. Murine epididymal cell lines were also derived from mice harboring a temperature-sensitive SV40 LTAg transgene [86]. Several other murine epididymal cell lines were derived from transgenic animals expressing an SV40 LTAg transgene under the control of the glutathione peroxidase-5 promoter [87]. Our laboratory has developed an immortalized rat epididymal principal cell line [88]. These cells (RCE-1) express transcripts for hormone receptors, epididymal-specific genes, and junctional proteins including several Cxs and Cldns. Immunofluorescence studies have revealed that these junctional proteins were localized to the points of cell–cell contact, indicative of functional intercellular junctions. More recently, we used a similar approach to develop human epididymal cell lines from both fertile and infertile patients [89, 90]. These cell lines form functional tight junctions and using siRNA, we demonstrated that downregulation of a single Cldn could alter the formation of tight junctions, as determined by measuring transepithelial resistance (TER) across the cells in culture. This indicated that a loss, or mistargeting of a single Cldn, as observed in nonobstructive infertile patients [89], is sufficient to compromise the blood–epididymis barrier. To further understand the role of Cldns and the blood–epididymis barrier in male infertility, we examined epididymides in infertile patients with obstructive azoospermia (OA).

We observed that mRNA levels of epididymal Cldns 1, 4, and 10 were altered down-stream from the obstruction site in OA patients. Epithelial cell lines were derived from the caput epididymidis of OA patients (infertile human caput epididymal cell line (IHCE)) [90]. IHCEs cells did not express many Cldns and were unable to form tight junctions. The use of cell lines has obvious limitations for immunolocalization studies since they comprise a single cell type, whereas the epididymis is a pseu-dostratified tissue with multiple cell types. Nevertheless, these cell lines have been particularly useful in elucidating cellular and molecular mechanisms. A recent study on the characterization of the ATP Binding Cassette B1 (ABCb1), also known as the multidrug resistance 1 transporter, used immunohistochemical techniques to characterize the cellular localization of ABCb1 in the rat epididymis [91]. This was complemented with studies using rat epididymal RCE cells to show that ABCb1 could be induced by alkylphenols, in this case nonylphenol and octylphenol, which could help explain the limited accumulation of alkylphenols in the epididymis of rats treated with octylphenol [92]. Thus, these cell lines from various animal models provide critical tools that can be used for toxicology studies.

Detoxification of Contaminants in Sperm

It is well accepted that spermatozoa are transcriptionally inert. Since DNA is com-pacted in haploid cells of spermatids during spermatogenesis, it is no longer capable of supporting transcription [93]. This limits the ability of maturing spermatozoa during epididymal transit to respond to chemical insult by modifying or stimulating gene expression to compensate, or metabolize, toxic chemicals. The most likely protective mechanism of the sperm resides in the basic architecture of the com-pacted DNA in its nucleus, which limits potential effects on the genetic material itself, given that the function of the sperm is to transport the DNA to the egg for fertilization. However, a second mechanism was recently reported, in which ABCB1 protein is expressed in epididymal spermatozoa. Immunolocalization of ABCB1 was confirmed using two different detection methods and further verified by west-ern blot. The mechanism by which this protein is expressed in the spermatozoa remains unknown; however, other proteins such as carnitine, an important biologi-cal molecule for sperm maturation and acquisition of motility, have also been local-ized to epididymal sperm [94]. Certain proteins, including both surface and GPI-anchored proteins, have been demonstrated previously to be transferred from the epithelium to sperm from vesicles secreted by epididymal principal cells and which are referred to as epididymosomes [95]. These are vesicles that are secreted be principal cells into the lumen and have been proposed as transporters of proteins to the maturing spermatozoa [95, 96]. Another possibility may be that a nonfunc-tional ABCB1 is present in testicular spermatozoa, and is unmasked during epididy-mal transit: ABCB1 is a phospho-glycoprotein with several posttranslational modification sites [97]. In support of this, Oko et al. [98] identified functional Golgi in the cytoplasmic droplet of epididymal spermatozoa. The Golgi may increase the glycosylation of ABCB1, allowing the protein to be recognized by the antibody. In any case, the presence of ABCB1 represents a unique mechanism by which spermatozoa may defend against certain classes of environmental toxicants.

Perspectives and Conclusions

There have been an increasing number of studies that have reported declining trends in male reproductive health over the past 40 years. These have included numerous studies on decreasing sperm counts, increased developmental malformations, testicular cancer, and decreasing levels of testosterone [5]. While there has not been any single irrefutable factor that has been shown to be responsible for this decline in reproductive health, the majority of studies suggest the environmental toxicants represent a major contributor to this decline. Although many contaminants cause overt pathological effects, the fact remains that environmental exposure to contaminants is likely to occur at very low levels over prolonged periods of time. As such, the effects of these contaminants are likely to cause more subtle cell-specific effects in either the testis and/or epididymis. Assessing these effects, therefore, requires histopathological approaches to determine whether or not cell functions are altered. In the past two decades, better detection systems and an increasing number of primary antibodies directed against a variety of key protein implicated in cell function have become available to localize proteins. Little doubt exists as to the increasing role and importance of immunolocalization of proteins in toxicological assessment of reproductive function.

References

1. Juul A, Main KM, Skakkebaek NE. Development: disorders of sex development-the tip of the iceberg? Nat Rev Endocrinol. 2011;7(9):504–5.
2. Matorras R, et al. Decline in human fertility rates with male age: a consequence of a decrease in male fecundity with aging? Gynecol Obstet Invest. 2011;71(4):229–35.
3. Welsh M, et al. Critical androgen-sensitive periods of rat penis and clitoris development. Int J Androl. 2010;33(1):e144–52.
4. Sharpe RM. Environmental/lifestyle effects on spermatogenesis. Philos Trans R Soc Lond B Biol Sci. 2010;365(1546):1697–712.
5. Sharpe RM, Skakkebaek NE. Testicular dysgenesis syndrome: mechanistic insights and potential new downstream effects. Fertil Steril. 2008;89(2 Suppl):e33–8.
6. Mathur PP, D'Cruz SC. The effect of environmental contaminants on testicular function. Asian J Androl. 2011;13(4):585–91.
7. Red RT, et al. Environmental toxicant exposure during pregnancy. Obstet Gynecol Surv. 2011;66(3):159–69.
8. Foster WG, Maharaj-Briceno S, Cyr DG. Dioxin-induced changes in epididymal sperm count and spermatogenesis. Environ Health Perspect. 2010;118(4):458–64.
9. Klinefelter GR. Actions of toxicants on the structure and function of the epididymis. In: Robaire B, Hinton BT, editors. The epididymis: from molecules to clinical practice. New York: Plenum; 2002. p. 353–70.

10. Hermo L, et al. Surfing the wave, cycle, life history, and genes/proteins expressed by testicular germ cells. Part 1: background to spermatogenesis, spermatogonia, and spermatocytes. Microsc Res Tech. 2010;73(4):241–78.
11. Hermo L, et al. Surfing the wave, cycle, life history, and genes/proteins expressed by testicular germ cells. Part 5: intercellular junctions and contacts between germs cells and Sertoli cells and their regulatory interactions, testicular cholesterol, and genes/proteins associated with more than one germ cell generation. Microsc Res Tech. 2010;73(4):409–94.
12. Dubé É, CYR DG. The blood-epididymis barrier. In: Cheng CY, editor. Blood-Tissue Barriers. Cambridge, MA: Landes Biosciences. 2012. pp 218–236.
13. Huckins C, Clermont Y. Evolution of gonocytes in the rat testis during late embryonic and early post-natal life. Archives d'anatomie, d'histologie et d'embryologie normales et experimentales. 1968;51(1):341–54.
14. Cooke PS. Thyroid hormones and testis development: a model system for increasing testis growth and sperm production. Ann N Y Acad Sci. 1991;637:122–32.
15. Bunick D, et al. Developmental expression of testis messenger ribonucleic acids in the rat following propylthiouracil-induced neonatal hypothyroidism. Biol Reprod. 1994;51(4):706–13.
16. Cooke PS, Zhao YD, Bunick D. Triiodothyronine inhibits proliferation and stimulates differentiation of cultured neonatal Sertoli cells: possible mechanism for increased adult testis weight and sperm production induced by neonatal goitrogen treatment. Biol Reprod. 1994;51(5):1000–5.
17. Van Haaster LH, et al. The effect of hypothyroidism on Sertoli cell proliferation and differentiation and hormone levels during testicular development in the rat. Endocrinology. 1992;131(3):1574–6.
18. van Haaster LH, et al. High neonatal triiodothyronine levels reduce the period of Sertoli cell proliferation and accelerate tubular lumen formation in the rat testis, and increase serum inhibin levels. Endocrinology. 1993;133(2):755–60.
19. Holsberger DR, et al. Cell-cycle inhibitors p27Kip1 and p21Cip1 regulate murine Sertoli cell proliferation. Biol Reprod. 2005;72(6):1429–36.
20. St-Pierre N, et al. Neonatal hypothyroidism alters the localization of gap junctional protein connexin 43 in the testis and messenger RNA levels in the epididymis of the rat. Biol Reprod. 2003;68(4):1232–40.
21. Brehm R, et al. A sertoli cell-specific knockout of connexin43 prevents initiation of spermatogenesis. Am J Pathol. 2007;171(1):19–31.
22. Sridharan S, et al. Proliferation of adult sertoli cells following conditional knockout of the Gap junctional protein GJA1 (connexin 43) in mice. Biol Reprod. 2007;76(5):804–12.
23. de Montgolfier B, Audet C, Cyr DG. Regulation of the connexin 43 promoter in the brook trout testis: role of the thyroid hormones and cAMP. Gen Comp Endocrinol. 2011;170(1):110–8.
24. Hermo L, et al. Surfing the wave, cycle, life history, and genes/proteins expressed by testicular germ cells. Part 2: changes in spermatid organelles associated with development of spermatozoa. Microsc Res Tech. 2010;73(4):279–319.
25. Leblond CP, Clermont Y. Definition of the stages of the cycle of the seminiferous epithelium in the rat. Ann N Y Acad Sci. 1952;55(4):548–73.
26. Russell LD. The blood-testis barrier and its formation relative to spermatocyte maturation in the adult rat: a lanthanum tracer study. Anat Rec. 1978;190(1):99–111.
27. Scheer H, Robaire B. Steroid delta 4-5 alpha-reductase and 3 alpha-hydroxysteroid dehydrogenase in the rat epididymis during development. Endocrinology. 1980;107(4):948–53.
28. Kleymenova E, et al. Exposure in utero to di(n-butyl) phthalate alters the vimentin cytoskeleton of fetal rat Sertoli cells and disrupts Sertoli cell-gonocyte contact. Biol Reprod. 2005;73(3):482–90.
29. Boekelheide K, et al. Dose-dependent effects on cell proliferation, seminiferous tubules, and male germ cells in the fetal rat testis following exposure to di(n-butyl) phthalate. Microsc Res Tech. 2009;72(8):629–38.
30. Yao PL, et al. Transcriptional regulation of FasL expression and participation of sTNF-alpha in response to sertoli cell injury. J Biol Chem. 2007;282(8):5420–31.

31. Moffit JS, et al. Mice lacking Raf kinase inhibitor protein-1 (RKIP-1) have altered sperm capacitation and reduced reproduction rates with a normal response to testicular injury. J Androl. 2007;28(6):883–90.
32. Kishta O, et al. In utero exposure to tributyltin chloride differentially alters male and female fetal gonad morphology and gene expression profiles in the Sprague-Dawley rat. Reprod Toxicol. 2007;23(1):1–11.
33. Pelletier RM. The blood-testis barrier: the junctional permeability, the proteins and the lipids. Prog Histochem Cytochem. 2011;46(2):49–127.
34. Gow A, et al. CNS myelin and sertoli cell tight junction strands are absent in Osp/claudin-11 null mice. Cell. 1999;99(6):649–59.
35. Meng J, et al. Androgens regulate the permeability of the blood-testis barrier. Proc Natl Acad Sci U S A. 2005;102(46):16696–700.
36. Cheng CY, et al. Environmental toxicants and male reproductive function. Spermatogenesis. 2011;1(1):2–13.
37. Cheng CY, Mruk DD. Cell junction dynamics in the testis: Sertoli-germ cell interactions and male contraceptive development. Physiol Rev. 2002;82(4):825–74.
38. Cyr DG, et al. Cellular interactions and the blood-epididymal barrier. In: Robaire B, Hinton BT, editors. The epididymis: from molecules to clinical practice. New York: Plenum; 2002. p. 103–18.
39. Cyr DG, Blaschuk OW, Robaire B. Identification and developmental regulation of cadherin messenger ribonucleic acids in the rat testis. Endocrinology. 1992;131(1):139–45.
40. Wu J, Jester Jr WF, Orth JM. Short-type PB-cadherin promotes survival of gonocytes and activates JAK-STAT signalling. Dev Biol. 2005;284(2):437–50.
41. Lin LH, DePhilip RM. Sex-dependent expression of placental (P)-cadherin during mouse gonadogenesis. Anat Rec. 1996;246(4):535–44.
42. Pospechova K, et al. Changes in the expression of P-cadherin in the normal, cryptorchid and busulphan-treated rat testis. Int J Androl. 2007;30(5):430–8.
43. Andersson AM, Edvardsen K, Skakkebaek NE. Expression and localization of N- and E-cadherin in the human testis and epididymis. Int J Androl. 1994,17(4).174–80.
44. Johnson KJ, Patel SR, Boekelheide K. Multiple cadherin superfamily members with unique expression profiles are produced in rat testis. Endocrinology. 2000;141(2):675–83.
45. Chung NP, Cheng CY. Is cadmium chloride-induced inter-sertoli tight junction permeability barrier disruption a suitable in vitro model to study the events of junction disassembly during spermatogenesis in the rat testis? Endocrinology. 2001;142(5):1878–88.
46. Siu ER, et al. An occludin-focal adhesion kinase (FAK) protein complex at the blood-testis barrier: a study using the cadmium model. Endocrinology. 2009;150(7):3336–44.
47. Cyr DG, Dufresne J, Gregory M. Cellular communication and the regulation of gap junctions in the epididymis. In: Hinton BT, Turner TT, editors. The third international conference on the epididymis. Charlottesville, VA: Van Doren; 2003. p. 50–9.
48. Plante I, Charbonneau M, Cyr DG. Decreased gap junctional intercellular communication in hexachlorobenzene-induced gender-specific hepatic tumor formation in the rat. Carcinogenesis. 2002;23(7):1243–9.
49. Plante I, Cyr DG, Charbonneau M. Sexual dimorphism in the regulation of liver connexin32 transcription in hexachlorobenzene-treated rats. Toxicol Sci. 2007;96(1):47–57.
50. Cyr DG, et al. Cellular immunolocalization of occludin during embryonic and postnatal development of the mouse testis and epididymis. Endocrinology. 1999;140(8):3815–25.
51. Carette D, et al. Major involvement of connexin 43 in seminiferous epithelial junction dynamics and male fertility. Dev Biol. 2010;346(1):54–67.
52. Li MW, et al. Connexin 43 is critical to maintain the homeostasis of the blood-testis barrier via its effects on tight junction reassembly. Proc Natl Acad Sci U S A. 2010;107(42): 17998–8003.
53. Li MW, et al. Connexin 43 and plakophilin-2 as a protein complex that regulates blood-testis barrier dynamics. Proc Natl Acad Sci U S A. 2009;106(25):10213–8.

54. Chung SS, Lee WM, Cheng CY. Study on the formation of specialized inter-Sertoli cell junctions in vitro. J Cell Physiol. 1999;181(2):258–72.
55. Risley MS, et al. Cell-, age- and stage-dependent distribution of connexin43 gap junctions in testes. J Cell Sci. 1992;103(Pt 1):81–96.
56. Aravindakshan J, et al. Consumption of xenoestrogen-contaminated fish during lactation alters adult male reproductive function. Toxicol Sci. 2004;81(1):179–89.
57. Aravindakshan J, Cyr DG. Nonylphenol alters connexin 43 levels and connexin 43 phosphorylation via an inhibition of the p38-mitogen-activated protein kinase pathway. Biol Reprod. 2005;72(5):1232–40.
58. Bedford JM. Changes in the electrophoretic properties of rabbit spermatozoa during passage through the epididymis. Nature. 1963;200:1178–80.
59. Orgebin-Crist MC. Sperm maturation in rabbit epididymis. Nature. 1967;216(5117):816–8.
60. Cyr DG, et al. Orchestration of occludins, claudins, catenins and cadherins as players involved in maintenance of the blood-epididymal barrier in animals and humans. Asian J Androl. 2007;9(4):463–75.
61. Turner TT. De Graaf's thread: the human epididymis. J Androl. 2008;29(3):237–50.
62. Hermo L, Robaire B. Epididymal cell types and their functions. In: Robaire B, Hinton BT, editors. The epididymis: from molecules to clinical practice. New York: Plenum; 2002. p. 81–102.
63. Veri JP, Hermo L, Robaire B. Immunocytochemical localization of the Yf subunit of glutathione S-transferase P shows regional variation in the staining of epithelial cells of the testis, efferent ducts, and epididymis of the male rat. J Androl. 1993;14(1):23–44.
64. Shum WW, et al. Transepithelial projections from basal cells are luminal sensors in pseudostratified epithelia. Cell. 2008;135(6):1108–17.
65. Yeung CH, et al. Basal cells of the human epididymis—antigenic and ultrastructural similarities to tissue-fixed macrophages. Biol Reprod. 1994;50(4):917–26.
66. Friend DS, Gilula NB. Variations in tight and gap junctions in mammalian tissues. J Cell Biol. 1972;53(3):758–76.
67. Pelletier RM. Freeze-fracture study of cell junctions in the epididymis and vas deferens of a seasonal breeder: the mink (Mustela vison). Microsc Res Tech. 1995;30(1):37–53.
68. Cyr DG, Robaire B, Hermo L. Structure and turnover of junctional complexes between principal cells of the rat epididymis. Microsc Res Tech. 1995;30(1):54–66.
69. Suzuki F, Nagano T. Development of tight junctions in the caput epididymal epithelium of the mouse. Dev Biol. 1978;63(2):321–34.
70. Gregory M, et al. Claudin-1 is not restricted to tight junctions in the rat epididymis. Endocrinology. 2001;142(2):854–63.
71. Gregory M, Cyr DG. Identification of multiple claudins in the rat epididymis. Mol Reprod Dev. 2006;73(5):580–8.
72. Cyr DG, Hermo L, Laird DW. Immunocytochemical localization and regulation of connexin43 in the adult rat epididymis. Endocrinology. 1996;137(4):1474–84.
73. Dufresne J, et al. Expression of multiple connexins in the rat epididymis indicates a complex regulation of gap junctional communication. Am J Physiol Cell Physiol. 2003;284(1):C33–43.
74. Cornwall GA, et al. Gene expression and epididymal function. In: Robaire B, Hinton BT, editors. The epididymis: from molecules to clinical practice. New York: Plenum; 2002. p. 169–200.
75. Cooper TG, et al. Gene and protein expression in the epididymis of infertile c-ros receptor tyrosine kinase-deficient mice. Biol Reprod. 2003;69(5):1750–62.
76. Cyr DG, et al. Expression and regulation of metallothioneins in the rat epididymis. J Androl. 2001;22(1):124–35.
77. Nonogaki T, et al. Localization of CuZn-superoxide dismutase in the human male genital organs. Hum Reprod. 1992;7(1):81–5.
78. Shum WW, et al. Establishment of cell-cell cross talk in the epididymis: control of luminal acidification. J Androl. 2011;32(6):576–86.

79. Ezner N, Robaire B. Androgenic regulation of the structure and functions of the epididymis. In: Robaire B, Hinton BT, editors. The epididymis: from molecules to clinical practice. New York: Plenum; 2002. p. 297–316.
80. Cyr DG, et al. Distribution and regulation of epithelial cadherin messenger ribonucleic acid and immunocytochemical localization of epithelial cadherin in the rat epididymis. Endocrinology. 1992;130(1):353–63.
81. Robaire B, Viger RS. Regulation of epididymal epithelial cell functions. Biol Reprod. 1995;52(2):226–36.
82. Cyr DG, Robaire B. Regulation of sulfated glycoprotein-2 (clusterin) messenger ribonucleic acid in the rat epididymis. Endocrinology. 1992;130(4):2160–6.
83. Hermo L, et al. Role of epithelial cells of the male excurrent duct system of the rat in the endocytosis or secretion of sulfated glycoprotein-2 (clusterin). Biol Reprod. 1991;44(6):1113–31.
84. Kaunisto K, et al. Regional expression and androgen regulation of carbonic anhydrase IV and II in the adult rat epididymis. Biol Reprod. 1999;61(6):1521–6.
85. Telgmann R, et al. Epididymal epithelium immortalized by simian virus 40 large T antigen: a model to study epididymal gene expression. Mol Hum Reprod. 2001;7(10):935–45.
86. Araki Y, et al. Immortalized epididymal cell lines from transgenic mice overexpressing temperature-sensitive simian virus 40 large T-antigen gene. J Androl. 2002;23(6):854–69.
87. Sipila P, et al. Immortalization of epididymal epithelium in transgenic mice expressing simian virus 40 T antigen: characterization of cell lines and regulation of the polyoma enhancer activator 3. Endocrinology. 2004;145(1):437–46.
88. Dufresne J, et al. Characterization of a novel rat epididymal cell line to study epididymal function. Endocrinology. 2005;146(11):4710–20.
89. Dube E, et al. Assessing the role of claudins in maintaining the integrity of epididymal tight junctions using novel human epididymal cell lines. Biol Reprod. 2010;82(6):1119–28.
90. Dube E, et al. Alterations in the human blood-epididymis barrier in obstructive azoospermia and the development of novel epididymal cell lines from infertile men. Biol Reprod. 2010;83(4):584–96.
91. Jones SR, Cyr DG. Regulation and characterization of the ATP-binding cassette transporter-B1 in the epididymis and epididymal spermatozoa of the rat. Toxicol Sci. 2011;119(2):369–79.
92. Hamelin G, et al. Determination of p-tert-octylphenol in blood and tissues by gas chromatography coupled with mass spectrometry. J Anal Toxicol. 2008;32(4):303–7.
93. Ortega MA, Sil P, Ward WS. Mammalian sperm chromatin as a model for chromatin function in DNA degradation and DNA replication. Syst Biol Reprod Med. 2011;57(1-2):43–9.
94. Kobayashi D, et al. Transport of carnitine and acetylcarnitine by carnitine/organic cation transporter (OCTN) 2 and OCTN3 into epididymal spermatozoa. Reproduction. 2007;134(5):651–8.
95. Sullivan R, Frenette G, Girouard J. Epididymosomes are involved in the acquisition of new sperm proteins during epididymal transit. Asian J Androl. 2007;9(4):483–91.
96. Sullivan R, et al. Role of exosomes in sperm maturation during the transit along the male reproductive tract. Blood Cells Mol Dis. 2005;35(1):1–10.
97. Juliano RL, Ling V. A surface glycoprotein modulating drug permeability in Chinese hamster ovary cell mutants. Biochim Biophys Acta. 1976;455(1):152–62.
98. Oko R, et al. The cytoplasmic droplet of rat epididymal spermatozoa contains saccular elements with Golgi characteristics. J Cell Biol. 1993;123(4):809–21.

Chapter 7
Immunohistochemistry and Female Reproductive Toxicology: The Ovary and Mammary Glands

Daniel G. Cyr, Patrick J. Devine, and Isabelle Plante

The Ovary

Oogenesis

Vertebrate ovarian follicles are composed of an oocyte surrounded by multiple layers of somatic cells. The inner most somatic cell layer is the granulosa, while the outer most layer forms the ovarian theca [1–3]. The process of oogenesis varies widely among different vertebrate groups, due in large part to different reproductive strategies between different species. In mammals, oogonia undergo rapid mitotic divisions during embryonic development. In humans, this begins a little after the midpoint of the first trimester and continues to the end of the second trimester, leading to the formation of approximately 7 million germ cells [1, 2]. After the seventh month of gestation, oogonia undergo extensive apoptosis and the remaining oogonia, referred to as primary oocytes, enter the first meiotic division to the diplotene stage. At this stage, the oocyte is enveloped by a single layer of cells referred to as the primordial follicle, consisting of granulosa and mesenchymal thecal cells [1, 3]. Groups, or clusters, of gametes at this stage are linked together by cytoplasmic bridges and undergo synchronous mitotic division which is likely regulated by factors exchanged between these cells via the cytoplasmic bridges [4]. Primary oocytes remain in the diplotene stage until puberty, when they resume meiosis following stimulation from blood-borne gonadotropins originating

D.G. Cyr (✉) • I. Plante
INRS-Institut Armand-Frappier, University of Quebec,
531 Boulevard des Prairies, Laval, QC, Canada H7B 1B7
e-mail: Daniel.Cyr@iaf.inrs.ca

P.J. Devine
Novartis Institutes for BioMedical Research, Inc., Cambridge, MA, USA

© Springer Science+Business Media New York 2016
S.A. Aziz, R. Mehta (eds.), *Technical Aspects of Toxicological Immunohistochemistry*, DOI 10.1007/978-1-4939-1516-3_7

from the adenohypophysis. Of the millions of primary oocytes, less than 400 undergo maturation during a typical woman's reproductive years [5, 6]. The resumption of meiosis by primary oocytes is accompanied by germinal vesicle breakdown, in which the integrity of the nuclear envelope breaks down, permitting the spindle fibers of metaphase to polarize to the extremities of the cell as the oocyte divides [3]. The second meiotic division, which forms the secondary oocyte, is accompanied by the formation of a smaller polar body. As primordial follicles undergo follicular growth, both the size of the oocyte and the layers of follicular cells dramatically increase [2]. The proliferation of follicular cells is regulated by the growth differentiation factor 9 (GDF9), a cytokine member of the TGF hormone family [7]. The innermost layer of the cells which surrounds the oocyte forms the cumulus of the oocyte. As the oocyte grows, its secretions will lead to the formation of a cavity, referred to as the antrum, which is composed of a variety of proteins, hormones, and other complex factors. Of the small group of oocytes that undergo maturation during a given menstrual cycle, most die and only a few mature to become tertiary, or Graafian, follicles [2, 3].

Data from transgenic mice have indicated essential roles of a number of genes during various phases of the follicular development. Pre-antral follicular growth is viewed to be gonadotropin independent, whereas antrum formation and follicular maturation require action of follicle-stimulating hormone (FSH). The forkhead transcription factor FOXL2 represents a major regulator of ovarian development, growth, and maturation [8]. It plays an important role early in ovarian development as well as in sex determination. Knockout studies have shown that in the absence of FOXL2, ovarian follicles fail to develop and result in partial sex reversal [9]. FOXL2 is also involved in folliculogenesis and is expressed in granulosa cells of small and medium follicles [10, 11]. FOXL2 acts as a transcriptional repressor of the mitochondrial cholesterol transporter StAR, as well as of CYP19, which codes for the P450 aromatase, thus inhibiting steroidogenesis in immature granulosa cells [11]. Several other genes are also downregulated by FOXL2, suggesting that its role may be a suppressor of folliculogenesis in small and medium granulosa cells. Similarly, b-cell leukemia/lymphoma-2 factor in the germ line α (FIGα) and GDF9 play critical role in pre-antral development [12–15].

Control of the initiation of follicle development of primordial follicles, termed follicle activation, involves many signals that are still not fully elucidated. The phosphatidylinositol 3-kinase (PI3K)/AKT, phosphatase, and tensin homolog deleted on chromosome ten (PTEN) signaling pathway is critical in controlling ovarian follicle activation. Oocyte-specific knockout in mice of either PTEN or FOXO3a, which are negative regulators of follicle activation, results in the activation of all primordial follicles and subsequent rapid follicle depletion and reproductive senescence [16, 17]. Furthermore, if a constitutively active FOXO3a is expressed in an oocyte-specific manner under the control of the zona pellucida 3 (ZP3) promoter, mice are infertile and follicle activation does not occur [18]. In contrast, if PTEN inactivation is placed under the control of the ZP3 promoter, which is activated in primary follicles, follicle development occurs normally and primordial follicles are activated at the same rate as in wild-type mice [19]. FOXO3a immunolocalization displays a nuclear localization in primordial follicles, mixed nuclear and cytoplasmic staining in primary follicles, and cytoplasmic staining in larger primary follicles [20].

FSH is critical for the maturation of follicles in forming tertiary or antral follicles [21]. As the ovarian follicle matures, its cells secrete a variety of growth factors that act on the oocyte to stimulate maturation as well as stimulating follicular development. Growth factors secreted by the oocyte, such as GDF9 and bone morphogenetic protein 15 (BMP15), also act on the developing follicles to stimulate granulosa cell proliferation [22]. Transgenic mice lacking the GDF9 gene are infertile and display early stage inhibition of follicular development [12]. Likewise, mutations in the BMP15 gene have been identified as the underlying cause of alterations in follicular growth and sterility observed in sheep [23, 24].

Assessing the effects of toxicants on the ovary using immunohistochemical approaches must, therefore, consider the unique development of the ovary as well as its regulation by various hormonal systems. Thus, some studies have examined the effects of chemicals such as TCDD on follicular granulosa cell proliferation using 5-bromo-2'-deoxyuridine (BrdU) incorporation [25]. BrdU is a synthetic nucleoside analogous to thymidine which is incorporated into DNA during cell proliferation. Using BrdU antisera, it is possible to immunolocalize BrdU in the nucleus and count BrdU-positive cells, and therefore evaluate cell proliferation. Alternatively, immunofluorescence for Ki67 has also been used to identify healthy follicles with proliferating granulosa cells [26, 27]. TUNEL (terminal deoxynucleotidyl transferase dUDP nick end labeling) staining has been used in a number of studies to identify atretic follicles [6]. Examples include small follicle destruction by the chemotherapeutic agent, cyclophosphamide [28] and DEHP [29]. In addition to using TUNNEL, the presence of pyknotic nuclei has also been used as indicators of follicular atresia [30]. Studies also examined the effect of toxicants on meiotic division using immunofluorescence and confocal microscopy. Bisphenol A, a plasticizer reported as having weak anti-estrogen, anti-androgen, and anti-thyroid hormone properties, has been shown in vitro to alter granulosa cell proliferation. Almost 20 % of oocytes exposed to bisphenol A failed to resume meiosis. Using α-tubulin to mark the spindles and ethidium homodimer-2 to mark the chromosomes, not only was meiotic division altered, but chromosomes were mis-aligned in almost 25 % of oocytes [31].

4-vinylcyclohexene and its diepoxide metabolite have been examined as a potential occupational exposure that could cause female infertility (NTP reports 362 (diepoxide, VCD) and 303 (4-vinylcyclohexene)). Although human exposures are not thought to occur at toxic levels, VCD is of interest for reproductive toxicology because it induces specific loss of primordial and primary follicles without affecting more developed follicles. Thus, it represents a good model for exposures which cause permanent ovarian failure. A role for activation of apoptotic signaling was suggested by increased activation of Caspase 3 specifically in the primordial and primary follicles (small) follicles targeted by VCD in rats as assessed by both Western blotting and immunofluorescence [32]. Furthermore, Caspase 8 and 9 protein levels increase in small follicles and mitochondria, respectively, following a single dose of VCD, but levels decrease in the cytosol after 15 days of treatment. Levels of proapoptotic Bad increased in both cytoplasmic and mitochondrial fraction following 15 days of VCD dosing, while BCL-x_L moved from the mitochondria to cytosolic fractions and BAX/BCL-x_L ratios increased, thereby suggesting proapoptotic signaling in response to VCD [33]. Immunohistochemical staining for cytochrome C, which is released from mitochondria during apoptosis, was increased in the cytoplasm of granulosa cells of small pre-antral follicles. Effects were not observed in larger follicle stages. Taken together, these results suggest that VCD induces apoptotic signaling leading to primordial and primary follicle loss.

Exposure to methoxychlor also results in atresia of pre-antral and antral follicles, respectively, through BAX-related mechanisms [30, 33–35]. Cigarette smoke has been reported to cause subfertility in females by decreasing the pool of primordial germ cells. Exposure decreases Bcl-2 but, unlike methoxychlor, does not induce apoptosis. In this case an increased number of autophagosomes in granulosa cells of ovarian follicles were noted. Furthermore, increased expression of Beclin-1 and microtubule-associated protein light chain 3, regulatory proteins that play a key role in the autophagy pathway, was observed in cigarette smoke-exposed mice [36].

Ovarian Endocrine System

In addition to various growth factors, the follicular cell layers work together in the synthesis of estrogens. Thecal cells are stimulated by LH to produce aromatizable androgens, testosterone, and androstenedione, which are converted to estradiol by aromatase (CYP19A1) in the granulosa cells [37].

Estrogens play critical roles in the menstrual cycle, influencing the thickening of the endometrium, thinning of the cervical mucus, and increasing FSH receptor levels on the granulosa cells, resulting in increased inhibin secretion which then acts on the pituitary to inhibit FSH secretion [37, 38]. As estradiol levels increase, there is a concomitant surge in LH levels that in turn stimulates ovulation of the fully mature oocyte [2].

Following ovulation, the luteal phase of the menstrual cycle begins. The follicle cells left behind after ovulation become the corpus luteum (Fig. 7.1) and, under the control of LH, secrete progestins that act on the lining of the uterus to prepare for implantation of the blastocyst. If the ovum is not fertilized, the corpus luteum degenerates, progesterone levels decrease, and the uterine lining is shed. As progesterone levels decrease, FSH levels begin to increase and a new cycle is initiated [1–3].

In addition to gonadotropins and steroid hormones, studies have shown that other hormones play an important role in the regulation of ovarian development. Thyroid hormones have received considerable attention over the years, although a clear understanding of thyroid hormones regulation remains somewhat elusive. In lower vertebrates such as fish, however, the role of the thyroid hormones on ovarian function is more clearly defined [39]. There have been numerous studies in which temporal profiles of thyroidal status and reproductive function have been compared

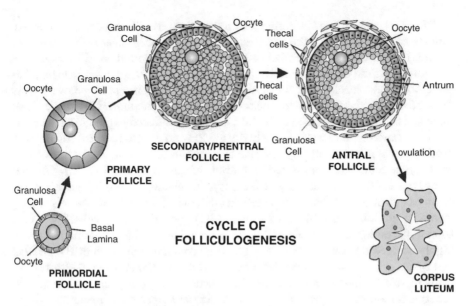

Fig. 7.1 Schematic diagram representing folliculogenesis in mammals. Primordial germ cells are characterized by having flattened granulosa along the basal lamina and surround the oocyte. Granulosa cells become larger and more cuboidal in primary follicles. The recruitment of thecal cells begins at the end of the primary follicle stage and a single layer surrounds the secondary or pre-antral follicle. At this stage the granulosa remain cuboidal. Secretions from the oocyte lead to the formation of a cavity, or antrum, which distinguishes the tertiary, or antral, follicle. This is also referred to as the Graafian follicle. At this stage, the thecal layer is vascularized and multiple layers of thecal cell cells are present. Following ovulation the remaining follicle degenerates into the corpus luteum and eventually into the corpus albicans

for many fish species of both genders. In many fish species, thyroid hormones play an important role in early oocyte development [39, 40]. In rainbow trout, triiodothyronine (T3) is particularly effective at low gonadotropin concentrations and appears to play a key role at the onset of gonadal development and subsequent reproduction [41, 42]. One component of thyroid hormone action involves cellular mechanisms blocking phosphodiesterase activity, thereby decreasing the rate of degradation of gonadotropin-stimulated intracellular cAMP [43]. Thyroid hormone levels also regulate the transactivation of the Cx43 promoter in vitro in the testis [44]. This effect is independent of cAMP, indicating that thyroid hormones can directly regulate gene expression in gonadal tissue [44].

In women, hypothyroidism has been associated with impaired ovulation, fertilization, implantation, miscarriage, and late pregnancy complications [45, 46]. In both hypo- and hyperthyroid women, there have been reports of alterations in the menstrual cycle [47]. These have generally been associated with effects on the LH/FSH cycles of these women [48]. A difficulty in establishing clinical interaction between altered thyroid function and infertility in women is the fact that most patients will likely be treated for their thyroid disorder before seeing a fertility specialist [45].

Direct actions of thyroid hormones on the ovary have also been documented. Studies have reported that the ovary expresses 5′-monodeiodinases (5′D) and can convert thyroxin (T4) into metabolically active T3 [49]. Both thyroid hormones are present in the fluid of antral follicles, and while levels of T4 are lower than circulating blood levels, levels of T3 are similar. 5′D activity is present in the ovary early during follicular development and peaks at the time of ovulation. Early studies have indicated that hypothyroidism in humans or in PTU-induced hypothyroid mice has been associated with decreased sensitivity to FSH and increased formation of polycystic ovary syndrome [50, 51]. Ovarian cyst formation has also been stimulated in hypothyroid animals administered high levels of gonadotropins [52, 53]. Whether or not this is due to an inability of the oocyte to mature or to other effects is not clear. Maruo et al. [54] have suggested that the role of thyroid hormones is associated with amplification of the FSH signaling on the ovarian thecal cells. These data support the notion that the role of thyroid hormone on ovarian maturation has been highly conserved in evolution. Given that many contaminants, such PCBs, PBDEs, and Bisphenol A, among others, have been shown to exert anti-thyroid effects [55], using thyroid hormone markers of ovarian function may be important when thyroid function is thought to represent a specific target for a given contaminant.

Cellular Communication in the Ovary

Cellular communication between cells occurs via endocrine, paracrine, or direct interactions. Direct cell–cell interaction results in the activation of signaling pathways in which information from the cell's external contacts is relayed into the cell to regulate gene expression in response to changing external contacts between cells. These interactions are mediated by cell adhesion molecules and tight junction proteins. Other signals are conveyed directly between the cytoplasms of adjacent cells via intercellular channels, referred to as gap junctions, that connect the cytoplasms of adjacent cells [56–58].

Gap junctions are comprised of a family of proteins that includes connexins, innexins, and pannexins [59]. Connexins and innexins are implicated in the formation of gap junctions between adjacent cells that are present in vertebrate and invertebrate species, respectively. Pannexins are also present in vertebrates and share homology with the innexins. However, unlike the innexins, pannexins do not form intercellular channels but rather cellular pores that are implicated in paracellular communication via the release of ATP [59]. Connexins are transmembrane proteins that oligomerize to form hemichannels or connexons [60, 61]. There are 21 different connexins named according to either their molecular mass or sequence homology. The connexons that form the gap junction can be composed of either the same connexin or of different connexins. The composition of the connexon will determine the charge of the pore and the types of molecules which are able to pass through the pore.

Cx43 IN RAT OVARIAN FOLLICLES

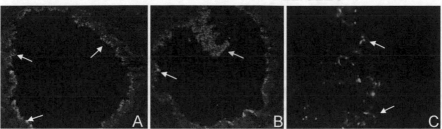

Fig. 7.2 Immunofluorescent localization of Cx43 in the oocyte of a rat ovary. Low power photomicrographs show extensive Cx43 immunostaining in the granulosa cells (*white arrows*). In the grazing section of *panel B*, one can observe the extensive reaction around the periphery of the cells suggestive of extensive gap junction formation (*yellow arrow*). At higher magnification (*panel C*), distinct punctate Cx43 immunostaining between adjacent follicular cells can be observed (*white arrows*). Nuclei were stained with propidium iodide and appear red while Cx43 is *green*

In all mammalian species studied to date, four to six different connexins are expressed in the ovary [56, 57]. In all species, Cx43 appears to be the most prevalent connexin and is expressed in granulosa cells (Fig. 7.2), while Cx37 is expressed by the oocyte [56]. Certain studies have suggested that Cx43 is also present between the oocyte and the cumulus/granulosa cell layer [62], although others have failed to observe Cx43 between these cell layers [63, 64]. Veitch [64] claimed that gap junctions between oocytes and granulosa cell layer in fact comprised Cx37. In support of this, Cx37 knockout mice exhibit arrested folliculogenesis at the early antral stage [63] and oocytes fail to become competent [63, 65]. The role of gap junctional communication between the oocyte and the granulosa cells may involve regulation of cAMP transfer. Inhibition of gap junctional communication using heptanol or carbenoxolone results in decreased levels of cAMP in the oocyte [66, 67].

The regulation of Cx43 in rodent granulosa cells is mediated by FSH/LH [68, 69]. FSH appears to play an important role early in follicular development and positively regulates the expression of Cx43 [68, 70]). However, prior to ovulation, when LH levels increase, there is a decrease in Cx43 [71]. This decrease appears related to the LH surge which inhibits Cx43 via the MAPK and PKA signaling pathways [72].

Cx43 knockout mice die shortly after birth [73]. These animals, however, have fewer oocytes at birth [74]. If the oocytes from Cx43 knockout mice are placed under the kidney capsule of a wild-type mouse, or cultured in vitro, folliculogenesis can proceed to the primary follicle stage, but subsequent development is impaired [74]. Similarly, in mice harboring an autosomal dominant point mutation within the Gja1 gene (*Gja1^{Jrt/+}* mice) which codes for Cx43, reduced granulosa cell coupling (0–20 % versus wild type) and impaired fertility are observed [75, 76]. This subfertility was due, at least in part, to impaired follicular development, with fewer pre-ovulatory follicles present in ovaries and reduced gonadotropin response. In addition to regulating folliculogenesis, Lin et al. (2003) have shown an inverse correlation between follicular apoptosis and Cx43 levels. Furthermore, the glucocorticoid dexamethasone, known to decrease apoptosis in ovarian follicles, increases levels of Cx43 [77, 78]. Interestingly, as Cx43 levels decrease in apoptosis, there is an increase in Cx32 levels. It has been suggested that Cx32 gap junctions propagate the death signal in the ovary [79, 80]. Thus, toxicants that stimulate a switch from Cx43 to Cx32 in the ovarian follicles may be useful for predicting ovarian apoptosis, thereby highlighting the importance of examining their expression and localization in toxicological studies. Whether or not toxicants like tributyltin (TBT), which induce apoptosis in the ovary during development, mediate changes in Cxs remains to be shown, even though TBT has been shown to decrease Cx43 in developing testes [81].

Toxicology data have also suggested a role of Cx43 in steroidogenesis. Exposure of rats to the environmental pesticide lindane blocks gap junctional communication [82]. In these animals there is also a concurrent decrease in the levels of P450ssc, the cholesterol transporter StAR, and a decrease in progesterone secretion, supporting the notion that Cx43 can influence steroidogenesis.

Ovary Cultures

In vitro culture of neonatal (postnatal day 0–4) rodent ovaries has been used to examine mechanisms underlying both ovarian follicular formation and development and toxicity of xenobiotics causing ovarian follicle loss [83–86]. Methodology involves isolating intact ovaries from young mice, rats, or hamsters and floating them on small pieces of nitrocellulose or other membrane which is in turn floating on the surface of culture medium [87, 88]. The majority of primordial follicles present remain dormant, but a slow steady number of follicles are able to develop normally to the secondary stage (multiple layers of granulosa cells) without added growth factors or hormones. Addition of growth factors or inhibitors can improve or inhibit follicle development or activation [86, 89]. Studies on direct toxic effects of drugs or pollutants have utilized this methodology as well. Follicles can be enumerated and staged by histology, gene expression can be monitored by either microarray or reverse transcription-polymerase chain reaction, and proteins can be localized by immunohistochemistry/immunofluorescence or quantified by Western blotting in spite of the small size of the tissue.

The role of DNA damage in the loss of primordial and primary follicles was examined, using cyclophosphamide as an example [28, 90]. Cyclophosphamide and other alkylating drugs induce DNA adducts, DNA double-strand cross-links, leading to DNA double-strand breaks [91]. In vivo studies determined that a reactive metabolite, phosphoramide mustard (PM), must be produced for follicle loss to occur [92]. Exposing cultured mouse ovaries to various concentrations of different metabolites of cyclophosphamide confirmed that only metabolites that could produce PM induced concentration-dependent loss of primordial and primary follicles [28]. TUNEL and cleaved Caspase 3 staining were used as markers of follicle injury or damage. Increased numbers of follicles with TUNEL staining were observed for 10 µM at 24 h and 3 µM at 48 h after exposure, suggesting that these markers are either not as sensitive as the follicle counts done 8 days after exposure or are time dependent, a more likely scenario. It was then determined whether or not DNA damage is an underlying factor in PM-induced ovarian follicle loss. Immunohistochemistry for phospho-histone H2AFX, a histone which becomes phosphorylated at sites of double-strand breaks [93], was performed to examine whether or not double-strand breaks can be observed in cultured rodent ovaries in response to PM [90]. Phospho-H2AFX staining was detected in a punctate pattern (foci) predominantly in oocytes of both cultured mouse and rat ovaries exposed to PM. Number of foci per oocyte and numbers of follicles with foci were both concentration dependent, peaking at 24–48 h with 3 and 10 µM. In another study, DNA cross-links in granulosa cells were observed and peaked 2 h following a single dose of CPA in immature Sprague–Dawley rats [94]. These results supported DNA damage as a mechanism of ovarian damage. Unfortunately, other immunohistochemical markers of DNA damage and repair remain to be developed for animal models, and more research is needed to further investigate oocyte-specific DNA damage signaling and repair capabilities in oocytes, as a unique target for DNA-damaging agents. In vitro work investigating VCD has examined the role of the KIT/KITL, PI3K, and FOXO3A signaling pathway in VCD-induced follicle loss. Initial evidence that follicle activation is affected by VCD was seen when a single dose of VCD induced an increase of primary follicles in rats [95]. Subsequent to that, global gene expression analyses of cultured rat ovaries and with isolated ovarian follicles identified cKit as a gene altered in both systems [96]. Addition of KITL, but not GDF9 or BMP4, to cultures partially protected against VCD-induced follicle loss. The PI3 kinase inhibitor, LY294002, inhibited follicle activation to primary follicles in cultured rat ovaries, and primordial follicles were not affected by VCD in the presence of LY294022, although numbers of small primary follicles were still reduced by VCD [97]. In cultures of postnatal day 4 rat ovaries, exposure to 30 µM VCD led to decreased immunofluorescence for KIT and FOXO3 protein in oocytes [98]. Decreased staining for phospho-AKT in oocyte nuclei was the earliest observed alteration on day 2 following VCD exposure.

Molecular Markers of Follicle Number and Health

Ovarian follicle counting is a necessary process for evaluating ovarian toxicants. Due to the relatively arduous process of determining follicle numbers during reproductive toxicology studies, automated or semi-automated methods have been sought. Typically for rodent studies, a thorough investigation of follicle numbers involves manually counting and determining the stage of all follicles in every 20th or 40th section of ovaries, following complete sectioning of one ovary per animal that has undergone standard processing and hematoxylin and eosin staining. Oocyte-specific markers were tested to either improve visibility of follicles for manual counting or for computer-assisted enumeration. PCNA has been tested, because oocyte nuclei stain darkly; however, proliferating granulosa cells are also stained and can be used to monitor granulosa cell proliferation and follicle health [99]. In spite of this lack of specificity, Muskhelishvili et al. [100] showed reduced variability in interindividual follicle counts with PCNA immunohistochemical staining. Furthermore, Picut et al. [101] demonstrated improved consistency and higher follicle numbers when PCNA-stained sections were counted manually, with a high correlation between manual counts and semi-automated computer-assisted counting. Cytochrome P450 1B1 immunostaining has also been reported to specifically stain oocyte nuclei and improve visibility of ovarian follicles for counting [102]. We have examined immunofluorescence of zona pellucida proteins (P Devine, Fig. 7.3, unpublished observations), using a specific antibody for this available from ATCC, originally developed by Jurrian Dean [103]. ZP3 are only present at the outer surface of oocytes in follicles of adults, including primordial oocytes although faintly (not shown). In adult

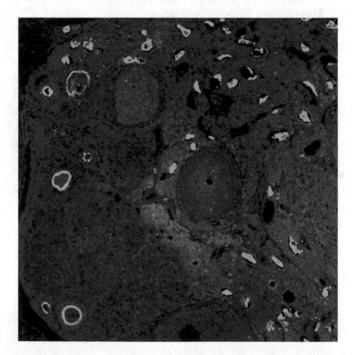

Fig. 7.3 Rat ovary stained for nuclei by Hoechst (*blue*) ZP3 (*green*) and anti-Mullerian hormone (*pink/red*). Corpora lutea had darker red background fluorescence, as well. ZP3 staining can be observed in structures that are likely follicle remnants, and can be differentiated from follicles by the shape of the structures and lack of oocytes within the zona pellucida

mice, ZP3 is also observed in follicle remnants, with the zona pellucida forming sinuous structures within the interstitium (P Devine, Fig. 7.3, unpublished observations). This would represent a potential fluorescent marker that could be used to enumerate and even stage follicles based upon oocyte size if the empty zona pellucidae could be removed from analysis. It could be envisaged that transgenic rodents expressing a fluorescent protein under the regulation of oocyte-specific genes could be developed for specific use in reproductive toxicology studies; however, the cost of such a model and the effort required to produce it have precluded such efforts.

In order to identify healthy versus atretic follicles, TUNEL staining or immunohistochemistry for cleaved Caspase 3 has been typically performed to identify apoptotic granulosa cells in atretic follicles [28, 104]. Conversely, immunohistochemistry for PCNA or BrDU incorporation of proliferating granulosa cells has been examined for healthy follicles with proliferating granulosa cells (e.g., [105]). Immunohistochemistry staining for Anti-Mullerian hormone has also been used to identify non-atretic antral follicles when cisplatin was compared with vehicle only [104]. These markers tend to be qualitative for follicle atresia or health, but they complement, or confirm, histological features such as apoptotic bodies or mitotic figures in secondary or antral follicles.

Markers of Ovarian Cancer

Ovarian tumors are generally rare in rodent species. We know from numerous studies, with different tissues, that changes in the levels of proteins implicated in cell–cell interactions can undergo dramatic changes in either tumorigenic cells or during the process of carcinogenesis. Intercellular gap junctional communication is known to be consistently downregulated in carcinogenesis, and this can be induced by chemical carcinogens. For example, in the liver of rats treated for 5 days with hexachlorobenzene (HCB), the expression of the gap junction proteins, Cx32 and Cx26, remains significantly lower 50 days after the beginning of the treatment, and GJIC is still decreased 100 days later, but only in female rats [106]. Interestingly, there is a preponderant increase in cancer formation in these female rats when they are subsequently given diethylnitrosamine (DEN), while connexins in the male are not affected by HCB and males develop significantly fewer tumors in response to DEN [106]. In a study using several ovarian cancer cell lines, most of the cell lines displayed reduced expression of connexins [107]. In granulosa cell tumors, it has been reported that Cx43 is decreased or absent. Interestingly, in cells with decreased Cx43 there is an increase in Cx32, which is normally absent in the normal ovary [108]. Thus, the regulation and interplay between these two connexins provide an interesting set of makers which may be useful in understanding chemical-induced carcinogenic transformation in the ovary. In another study, cisplatin treatment of ovarian cancer cells resulted in the formation of both sensitive and resistant cells. Downregulation of p53 in both cell types with siRNA resulted in downregulation of several genes including Cx43. In fact, cells in which Cx43 expression was decreased showed increased drug resistance to cisplatin [109]. Whether or not drugs that induce transporters, such as those of the ABC transporter family, have lesser effects in cells when Cx43 is downregulated remains to be established. However, dual labeling of connexins and certain known transporters may provide important information on potential drug/chemical toxicity in the ovary.

FSH receptor knockout mice (FORKO) develop ovarian tumors after 12 months of age [110]. These animals are hypergonadotropic but display hypogonadism and have been used as an animal model for menopause. After 12 months of age, FORKO mice develop different types of ovarian tumors, including granulosa cell tumors and serous papillary epithelial adenomas [110]. Ovaries of FORKO mice show increased levels of the tight junction protein claudins (Cldns) 3 and 4, as well as dramatic increases in Cldn11; Cldn1, on the other hand, is repressed [111]. These animals also display steroid hormone imbalance, with high levels of testosterone and low levels of progesterone [111]. These observations are important with regard to identifying effects associated with long-term exposure to steroid-mimicking endocrine-disrupting chemicals, particularly putative androgenic chemicals such as phthalates [112], or other endocrine-disrupting chemicals that alter the production of gonadotropins. Endocrine-disrupting chemicals are molecules that can mimic or inhibit the action of endogenous hormones on target tissues, or alter either hormone secretion or the synthesis of their receptors [113]. Thus, monitoring the expression levels and immunolocalization of ovarian Cldns or connexins represent strong histological biomarkers to evaluate the effects of these chemicals and whose changes may be associated with infertility or rendering the ovary to a precancerous state.

Toxicants and Mammary Gland Development

Over the past 30 years, many studies have demonstrated that mammary glands develop earlier in women from many countries, as compared to the 1930s–1940s [114–117]. Estrogens, secreted by the ovaries under the influence of the follicle-stimulating hormone (FSH) and the luteinizing hormone (LH), are essential for mammary gland development during puberty [118, 119]. However, precocious breast development is not associated with an earlier menarche, implying an action of estrogens without the activation of the hypothalamic–pituitary axis [120]. It has been suggested that environmental factors mimicking estrogens actions may be responsible for those perturbations [120–122]. Because endocrine disruptors are widespread in the environment, and because their effects at low concentrations can have major impacts on the tightly orchestrated hormone-dependent development of mammary gland, exposure to endocrine disruptors is particularly disconcerting for women [113, 123, 124]. Mice models are typically used to assess developmental defects induced by chemical exposure or genetic modifications on mammary gland development. While whole mounts have been used to visualize gross developmental defects in mammary glands, the recent development of molecular markers, as well as new in vitro models, allows the assessment of subtle changes at the cellular and molecular levels, particularly changes in cell–cell interactions.

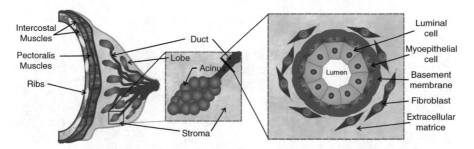

Fig. 7.4   Schematic diagram of the anatomical structure of the mammary gland

Structure of the Mammary Gland

The mammary gland is formed by two compartments, the stroma (or parenchyma) and the epithelium (Fig. 7.4). The stroma physically and nutritionally supports the epithelium [125]. It is composed mainly of adipocytes and fibroblasts but also contains endothelial cells and cells of the immune system, all of which influence mammary gland development [126]. In humans, the mature epithelium resembles a tree-like branching system comprising 15–20 different lobes. These lobes are themselves formed by numerous terminal lobular units of secretory alveoli (also referred to as acini) and converging ducts (Fig. 7.4). The epithelium consists of two layers of cells: an inner layer of luminal cells surrounded by an outer layer of myoepithelial cells (Fig. 7.4), and basement membrane [127, 128]. These structures develop at different stages of mammary gland organogenesis, and result in a substantial remodeling of the gland induced primarily by hormones. Since the structure and development of the murine mammary gland is similar to that of the human, this aspect of the review will focus on the murine mammary gland model.

Mouse Mammary Gland Development

Embryonic Development

Mammary gland development in mice begins around embryonic day 10.5 with the formation of bilateral milk lines (Fig 7.5a) located on the abdomen of the animal [129]. These milk lines are produced by epidermal cells which become columnar and multilayered. By day E11.5, cells within the milk line migrate to specific locations to form five pairs of lens-shaped placodes (Fig. 7.5a) [130]. During the following days, placodes enlarge and invaginate to form bulb-shaped buds, while cells located directly around the developing buds condense and differentiate, to form the primary mammary gland mesenchyme. The final phase of embryonic development begins around day E15.5, with proliferation of epithelial cells and elongation of the

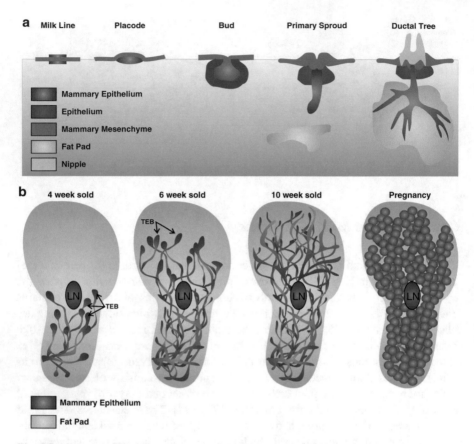

Fig. 7.5 Mammary Gland Organogenesis. Schematic model depicting (**a**) embryonic mammary gland development, (**b**) ductal elongation during puberty and alveologenesis during pregnancy. *TEB* Terminal end bud, *LN* Lymph Node

primary sprout into the mesenchyme. Simultaneously, mesenchymal cells located at the neck of the bud and specialized epithelial cells form the nipple at the surface of the body. Once the growing sprout has reached the developing fat pad under the dermis, it gives rise to the rudimentary ductal tree that is present at birth (Fig. 7.5).

Ductal Elongation

After birth, the mammary gland remains quiescent until the beginning of puberty (~4 weeks old), when the rise in circulating hormones and growth factors induces epithelial cell proliferation (Figs 7.5b and 7.7). The ductal elongation occurs in highly proliferating specialized structures located at the tips of the ducts called terminal end buds (TEB) [131]. It has been demonstrated that estrogens and estrogen receptor α (ERα) are required for ductal elongation, since ERα knockout mice

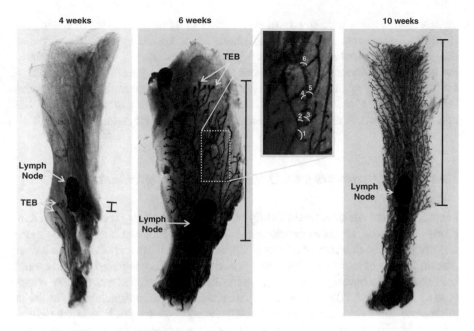

4 weeks   6 weeks   10 weeks

Fig. 7.6 Mammary Gland Whole Mount. Representative pictures of mammary gland whole mounts at 4, 6, and 10 weeks of age showing the lymph node and terminal end buds (*TEB*). *Bars* on the *right-hand side* represent the extent of ductal extensions from the *bottom* of the lymph node. *Inset* demonstrates how ramifications can be counted in order to evaluate mammary gland development

display impaired mammary gland development [132–134]. Similarly, the epithelial growth factor (EGF) and the insulin growth factor-1 (IGF-1) are necessary for proper ductal elongation [119]. At the end of puberty, around week 10 (Fig. 7.5b and 7.6), a complex tree-like system of ducts exists throughout the entire stroma and TEBs are no longer present [127]. The epithelium then undergoes cycles of branching and regression with each estrous cycle [127]. Circulating levels of progesterone and progesterone receptor (PR) are necessary for the formation of these tertiary side branches along the ducts. In some strains of mice, lobulo-alveolar structures also develop at the ends of tertiary branches [135].

Alveologenesis, Lactation, and Involution

At the end of puberty, the mammary gland is not functional. Function, namely milk production and lactation, is gained after another phase of remodeling induced by hormonal changes at the onset of pregnancy [136]. Epithelial cells proliferate and differentiate to form grape-shaped structures, the alveoli, where milk is produced and stored (Fig. 7.5b). Luminal cells inside the alveoli become secretory, while myoepithelial cell bodies and processes extend around the alveoli to form a basket-shaped network [52]. Expression of milk proteins and cell survival and pro-liferation signals required for alveologenesis are triggered by increased levels of progesterone and prolactin [137]. During lactation, suckling of the pups stimulates the secretion of oxytocin, which induces contraction of myoepithelial cells around the alveoli and along the ducts to transport milk [138]. After weaning, milk accumu-lation within the alveoli induces a cascade of events resulting in massive epithelial cell apoptosis, which returns the mammary gland back to its mature stage [139].

Endocrine Disruptors and Mammary Gland Development

Mammary gland development is tightly regulated and requires precise levels of hor-mones at different stages of development. Mammary cells can, therefore, be more susceptible to the detrimental effects of endocrine disruptors and other chemicals at certain periods of exposure. Even though hormones are not required for embryonic mammary gland development [119], studies have demonstrated that in utero expo-sure to bisphenol A (BPA), genistein, phthalates, atrazine, or vinclozolin results in precocious embryonic development [140], in addition to developmental defects observed during puberty and in adults. These include changes in morphology of the gland, cell proliferation, and gene expression [141–145]. Similarly, perinatal expo-sure to these chemicals results in increased numbers of both TEB at puberty [146] and alveolar buds in adult mice [147]. Pubertal ductal development involves cell proliferation, differentiation, and apoptosis, all of which are induced by increasing levels of steroid hormones [119]. This period, therefore, is also sensitive to develop-mental defects induced by exposure to endocrine disruptors [148]. For instance, rat exposed to BPA, diethylstilbestrol (DES), genistein, ep-DDT, Aroclor 1221, Aroclor 1254, or 2,3,7,8-tetrachlorodibenzo-p-dioxin (TCDD) during puberty showed higher than normal rates of mammary epithelial cell proliferation and premature development of lobular terminal ductal structures [149, 150]. Alveologenesis during pregnancy is another period of hormone-induced epithelial cell proliferation and differentiation that can be disturbed by endocrine disruptors. While few toxicologi-cal studies have focused on this period of development, it has been reported that exposure to dioxin during pregnancy resulted in poor lobulo-alveolar development and impaired lactation in mice [151]. Similarly, data from genetically modified mice demonstrated that alveologenesis, milk secretion, and lactation were impaired by improper signaling through hormone receptors [152], suggesting that exposure to other endocrine disruptors could result in similar defects.

Together, these data support the notion that exposure to endocrine disruptors, and in particular to those that mimic estrogenic effects, can promote precocious breast develop-ment and may contribute to the other effects observed in humans [120–122]. Moreover, the fact that embryonic and perinatal periods of mammary gland development appear to be sensitive to EDs is of particular concern, since many endocrine disruptors are found at higher concentrations in infants and children than in adults [153–156]. Over the years, various methods and models have been developed which can now be used to understand deleterious effects of endocrine disruptors EDs on mammary gland development.

Methods to Study Mammary Gland Development, Function, and Toxicology

One of the classical, and most commonly used, methods to study mammary gland developmental defect is whole mount. Generally, ductal development is assessed by evaluating the number of TEBs, ducts, and sub-branches and the length of ducts during embryonic development, during puberty, and in adults (Fig. 7.6) [157–161]. While alveologenesis can also be evaluated by whole mount, quantification during late pregnancy or lactation stages is difficult due to the density of the tissue. Moreover, whole mount gives only a macroscopic evaluation of the mammary gland structures. As a result, improved histological and imaging technologies are now used to evaluate defects at the molecular and cellular levels.

Imaging the Intercellular Junctions in Mammary Glands

The need for interaction between the epithelium and the stroma to propagate signals generated from hormonal responses was confirmed years ago by numerous transplant and in vitro studies [162–167]. More recently, studies have demonstrated that mammary gland development and function require bridging and connecting of the myoepithelial and luminal cells, a function ensured by gap, adherens, and tight junctions (Fig. 7.5a) [168–171]. As indicated previously in the text, gap junctions are channels formed by connexins, which permit direct communication between adjacent cells [172]. Adherens junctions are composed of transmembrane and cytoplasmic proteins, including cadherins and catenins, and are involved in adhesion between cells and in cell signaling [173]. Tight junctions are composed of transmembrane proteins, claudins, occludin, and tricellulin, bound to a complex of cytoplasmic proteins [174]. Tight junctions form a semipermeable barrier that controls paracellular transport between the cells and allows for cell polarity [173–175]. Studies using genetically modified mice models have shown that proteins forming gap, tight, and adherens junctions are required for proper mammary gland development [158, 176–178]. Moreover, it has been shown that many endocrine disruptors can modulate junctional protein expression in various tissues, resulting in developmental defects and cancer promotion [106, 179–185]. Thus, imaging of junctional proteins represents a significant tool for assessing mammary gland development in toxicological studies.

Luminal or myoepithelial localization of junctional proteins has been traditionally assessed by immunohistochemistry [186, 187] or electron microscopy [171, 188]. However, the increased availability of high-quality antibodies and accessibility to confocal microscopes now allow for easier imaging of the mammary gland components. As with many multilayered epithelia, luminal and myoepithelial cells express different types of cytokeratins (cytokeratins 8/18 (K18) and 14 (K14), respectively) and other markers facilitating the localization of proteins (Fig. 7.7b and c) [189]. Consequently, we and others have demonstrated that in the mouse

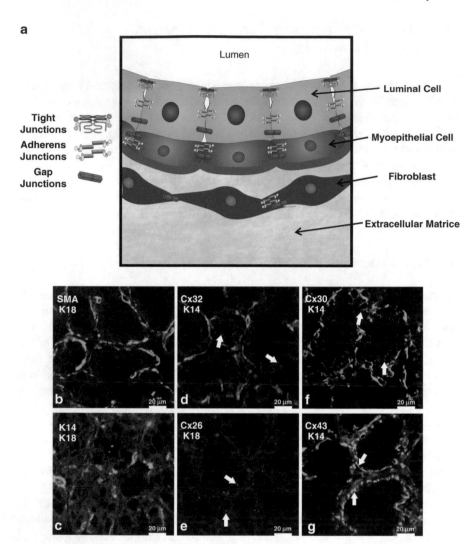

Fig. 7.7 Localization of intercellular junctions in mammary gland structure. (**a**) Schematic model depicting the localization of tight, adherens, and gap junctions in mammary gland. (**b**) Representative sections of mice mammary glands at parturition immunolabeled for luminal cell-specific keratin 18 (K18, *red*; **b, c, e**), myoepithelial cell-specific keratin 14 (K14, *green*; **c, d, f, g**), or Smooth muscle actin (SMA, *green*; **b**). Cx32 (**d**), Cx26 (**e**), and Cx30 (**f**) punctate structures typical of gap junction plaques were present at the cell surface of luminal cells while Cx43 (**g**) (*purple*, **d–g**) was present at the surface of myoepithelial cells

mammary gland, Cx26, Cx30, and Cx32 are expressed at the plasma membrane of luminal cells, and expression levels of these junctional proteins vary throughout development (Fig. 7.7d–f) [190–192]. Similarly, Cx43 is expressed at the plasma membrane mainly of K14-positive myoepithelial cells and of fibroblasts at all stages of mammary gland development (Fig. 7.7 g) [158, 191].

Cellular junctions are also required for proper function of the mammary gland. Tight junctions, in particular, allow the formation of a lumen by forming a barrier between the milk and the interstitial fluid [188, 193]. Moreover, tight junction closure is required for secretory activation. While imaging technologies can be used to determine proper localization of junctional proteins, tight junction function needs to be assessed by measuring transepithelial resistance [194] or by intraductal injection of a foreign substance, such as [^{14}C]-sucrose into the lumen of the gland [188]. The presence of cytoplasmic lipid droplets (CLD) after parturition is another indication of improper activation of secretion. While the presence of CLDs is normal during late pregnancy, CLDS are typically replaced by small lipid droplets at the apical surface of the epithelial cells after parturition [195]. CLDs can be observed in histological sections (Fig. 7.8a, b) and quantified using adipophillin immunostaining, which coats the surface of CLDs and milk lipid globules [196]. Another phenotype associated with a leaky epithelium or impaired mammary gland function is the presence of milk in the lumen of the alveoli after parturition. To quantify and determine if milk is more frequently found in the alveoli lumens of mammary glands from mutant mice or mice exposed to endocrine disruptors, histological sections of mammary glands can be double stained with a milk protein, such as whey acidic protein (WAP), and lectin wheat germ agglutinin (WGA) (Fig. 7.8c, d). WGA binds to the apical border of the luminal cells, providing a visual indicator of the luminal border of alveoli and ducts. Mis-localization of WGA or increased detection of milk components throughout the lumen can thus reflect developmental or functional defects (Fig. 7.8) [191]. An elegant study recently published suggested that developmental and functional defects can be assessed at different stages of mammary gland organogenesis at the molecular levels using an array of markers referred to as "tissue proteotyping" [197].

In Vitro Models for Studying Mammary Glands

A limitation of immunohistochemistry is the ability to visualize the structure of the mammary gland during alveologenesis. While myoepithelial cells surrounding luminal cells in ducts are oriented parallel to the long axis of the ducts, forming a continuous layer that is easily identifiable (Fig. 7.9a), myoepithelial cell bodies and processes extend around the alveoli during pregnancy and lactation, forming a basket-shaped network (Fig. 7.9b) [52]. As a result of these structural changes, developmental defects induced by toxicants or resulting from mutations can be easily observed in luminal cells, but may be undetectable in myoepithelial cells.

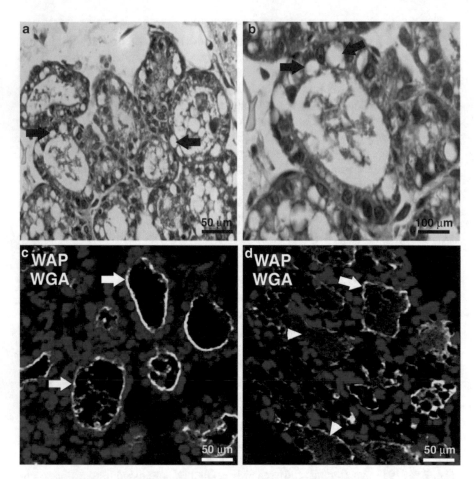

Fig. 7.8 Evaluation of secretion activation and lactation defects using imaging. (**a, b**) The presence of cytoplasmic lipids droplets (*arrows*) in histological sections after parturition is an indication of improper secretion activation. Paraffin-embedded mammary gland sections from mice after parturition were stained with eosin and hematoxylin. (**c, d**) Paraffin-embedded mammary gland sections were labeled with wheat germ agglutinin (WGA, *green*) to denote the luminal border (*arrows*) and anti-whey acidic protein (WAP) antibody (*red*) to indicate the presence of milk proteins (*arrowheads*)

Fig. 7.9 (continued) Nuclei were stained with Hoechst (*blue*). (**c**) Cells isolated from mammary gland were mainly of myoepithelial origin, as confirmed by keratin 14 staining, and were competent in passing Lucifer yellow to neighboring cells (**d**), an indication of gap junctional intercellular communication. The *asterisks* indicated which cell was microinjected with Lucifer yellow dye. (**e**) Western blot analysis indicated that connexin 43 (Cx43), β-catenin (β-cat), and Smooth Muscle Actin (SMA) were expressed in primary myoepithelial cells. Protein levels were normalized to GAPDH. (F-I) MCF12A (luminal) were cultured (**f, g**) or cocultured with primary myoepithelial cells (**h, i**) in Matrigel for 7 days. Light microscopy (**f, h**) and confocal images (**g, i**) show acini formation by MCF12A (**f, g**) and organotypic structures (**h, i**). Cells were immunolabeled for either β-catenin (*purple*, **g**) or keratin 14 (*green*, **i**). Nuclei were stained with Draq5 (*red*). **i** is superimposed on its own DIC image

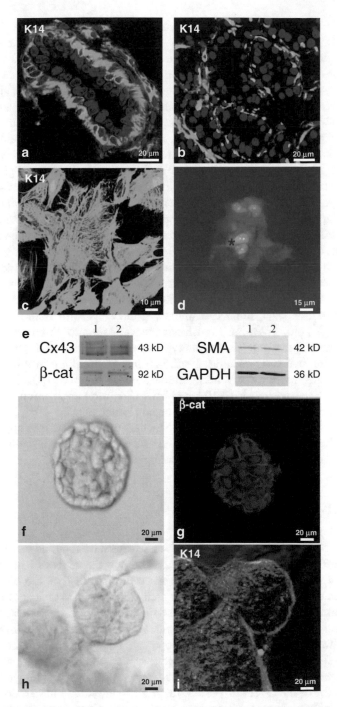

Fig. 7.9 In vitro models to study mammary gland. (**a**, **b**) Representative sections of mouse mammary glands at 10 weeks of age (**a**) or at parturition (**b**) immunolabeled for keratin 14 (K14, *green*), a marker of myoepithelial cells, showing the shape and localization of the myoepithelial cells.

Moreover, while animal studies are crucial to evaluate the toxicological effects of compounds, they are both cost- and time-demanding. To overcome the limitations of animal studies, many mammary luminal cell lines have been used for in vitro toxicity assays, with excellent results. On the other hand, myoepithelial cells have been mostly ignored, both in developmental and toxicological studies, since it was believed that their role was limited to milk transport and ejection during lactation. We now know that myoepithelial cells are required for proper mammary gland differentiation and function [52, 198–200]. As there are few tools available to study myoepithelial cells, we have developed a new method for the isolation and culture of pure populations of myoepithelial cells from murine mammary glands [158, 191]. Cultured myoepithelial cells are differentiated, as they retain expression of specific myoepithelial markers, and these maintain the capacity to communicate via gap junctions and to contract upon oxytocin exposure (Fig. 7.9c–e). Primary myoepithelial cells offer a new avenue in toxicological studies to understand the effects of toxicants on these important cells.

Although 2D culture systems are extensively used and are valid models, they do not mimic the 3D structure of organs. Recent studies have used 3D models in which cells are cultured within a matrix containing components of the extracellular matrix. When cultured as such, luminal mammary gland cells form acini-like structures (Fig. 7.7f, g) [180, 184]. An even more innovative approach can also be used, namely, "organotypic" models. Organotypic models consist of a mixture of myoepithelial and luminal cells cocultured within a matrix, resulting in 2-layered acini, thus mimicking more closely the mammary gland structure [201–203]. Human luminal cell lines, such as MCF10A and MCF12A, form polarized double-layered acini when cocultured with primary myoepithelial cells (Fig. 7.9h, i). Combined with imaging techniques, these new models thus provide a unique opportunity to study the effects of toxicants not only on luminal and myoepithelial cells but also on the junctional interactions between the two epithelial cell layers.

Perspectives and Conclusions

There have been an increasing number of studies that have reported increased use of in vitro fertilization throughout the world. Through the advent of newer in vitro fertilization technologies, many women are waiting until they are older to have babies, thus extending potential exposure period of oocytes to toxicants. It is, therefore, important to elucidate the cellular and molecular processes which occur in the reproductive system. Recent findings provide novel approaches to explore how contaminants may alter cell gonadal development. Furthermore, because the ovary undergoes substantial changes during oogenesis, and because it comprises different cell types, the use of immunolocalization is essential for understanding pathological changes in the ovary. Over the past several years, many studies have reported higher incidences of breast cancer and precocious mammary gland development in women from many countries. Since mammary gland development is highly regulated by steroid and hypophyseal hormones, slight changes in hormonal balance can result in alterations to both development and function. Whether or not precocious mammary gland development contributes to increased chance of breast cancer development later in life, or if the chemicals that stimulate mammary gland development predispose the gland to the development of cancer, remains to be fully addressed. The use of sensitive tools such as markers of cell–cell interactions in both animal and human tissue samples, as well as in vitro cell lines and 3D cell culture models, represents important components of future screening programs aimed at understanding effects of toxicants on both ovary and mammary glands.

Acknowledgments The authors wish to thank Mary Gregory (INRS-Institut Armand-Frappier) for her suggestions and immunolabeling of rat ovaries. The authors also wish to acknowledge research funding from NSERC and the Réseau de recherches en santé environnementale du FRSQ.

References

1. Fortune JE. Ovarian follicular growth and development in mammals. Biol Reprod. 1994; 50:225–32.
2. Gilbert SF. Developmental biology. 6th ed. Sunderland, MA: Sinauer Associates; 2000.
3. Hirshfield AN. Overview of ovarian follicular development: considerations for the toxicologist. Environ Mol Mutagen. 1997;29:10–5.
4. Merchant H, Zamboni L. Presence of connections between follicles in juvenile mouse ovaries. Am J Anat. 1972;134:127–32.
5. Gougeon A. Regulation of ovarian follicular development in primates: facts and hypotheses. Endocr Rev. 1996;17:121–55.
6. Krysko DV, Diez-Fraile A, Criel G, Svistunov AA, Vandenabeele P, D'Herde K. Life and death of female gametes during oogenesis and folliculogenesis. Apoptosis. 2008;13: 1065–87.
7. Pangas SA, Matzuk MM. Genetic models for transforming growth factor beta superfamily signaling in ovarian follicle development. Mol Cell Endocrinol. 2004;225:83–91.
8. Pisarska MD, Barlow G, Kuo FT. Minireview: roles of the forkhead transcription factor FOXL2 in granulosa cell biology and pathology. Endocrinology. 2011;152:1199–208.
9. Schmidt D, Ovitt CE, Anlag K, Fehsenfeld S, Gredsted L, Treier AC, Treier M. The murine winged-helix transcription factor Foxl2 is required for granulosa cell differentiation and ovary maintenance. Development. 2004;131:933–42.
10. Ottolenghi C, Omari S, Garcia-Ortiz JE, Uda M, Crisponi L, Forabosco A, Pilia G, Schlessinger D. Foxl2 is required for commitment to ovary differentiation. Hum Mol Genet. 2005;14:2053–62.
11. Pisarska MD, Bae J, Klein C, Hsueh AJ. Forkhead l2 is expressed in the ovary and represses the promoter activity of the steroidogenic acute regulatory gene. Endocrinology. 2004;145: 3424–33.
12. Dong J, Albertini DF, Nishimori K, Kumar TR, Lu N, Matzuk MM. Growth differentiation factor-9 is required during early ovarian folliculogenesis. Nature. 1996;383:531–5.
13. Greenfeld CR, Babus JK, Furth PA, Marion S, Hoyer PB, Flaws JA. BAX is involved in regulating follicular growth, but is dispensable for follicle atresia in adult mouse ovaries. Reproduction. 2007;133:107–16.
14. Hsu SY, Lai RJ, Finegold M, Hsueh AJ. Targeted overexpression of Bcl-2 in ovaries of transgenic mice leads to decreased follicle apoptosis, enhanced folliculogenesis, and increased germ cell tumorigenesis. Endocrinology. 1996;137:4837–43.

15. Soyal SM, Amleh A, Dean J. FIGalpha, a germ cell-specific transcription factor required for ovarian follicle formation. Development. 2000;127:4645–54.
16. Castrillon DH, Miao L, Kollipara R, Horner JW, DePinho RA. Suppression of ovarian follicle activation in mice by the transcription factor Foxo3a. Science. 2003;301:215–8.
17. Reddy P, Liu L, Adhikari D, Jagarlamudi K, Rajareddy S, Shen Y, Du C, Tang W, Hamalainen T, Peng SL, Lan ZJ, Cooney AJ, Huhtaniemi I, Liu K. Oocyte-specific deletion of Pten causes premature activation of the primordial follicle pool. Science. 2008;319:611–3.
18. Liu L, Rajareddy S, Reddy P, Du C, Jagarlamudi K, Shen Y, Gunnarsson D, Selstam G, Boman K, Liu K. Infertility caused by retardation of follicular development in mice with oocyte-specific expression of Foxo3a. Development. 2007;134:199–209.
19. Jagarlamudi K, Liu L, Adhikari D, Reddy P, Idahl A, Ottander U, Lundin E, Liu K. Oocyte-specific deletion of Pten in mice reveals a stage-specific function of PTEN/PI3K signaling in oocytes in controlling follicular activation. PLoS One. 2009;4, e6186.
20. John GB, Shirley LJ, Gallardo TD, Castrillon DH. Specificity of the requirement for Foxo3 in primordial follicle activation. Reproduction. 2007;133:855–63.
21. Kumar TR, Wang Y, Lu N, Matzuk MM. Follicle stimulating hormone is required for ovarian follicle maturation but not male fertility. Nat Genet. 1997;15:201–4.
22. Juengel JL, Bodensteiner KJ, Heath DA, Hudson NL, Moeller CL, Smith P, Galloway SM, Davis GH, Sawyer HR, McNatty KP. Physiology of GDF9 and BMP15 signalling molecules. Anim Reprod Sci. 2004;82–83:447–60.
23. Galloway SM, McNatty KP, Cambridge LM, Laitinen MP, Juengel JL, Jokiranta TS, McLaren RJ, Luiro K, Dodds KG, Montgomery GW, Beattie AE, Davis GH, Ritvos O. Mutations in an oocyte-derived growth factor gene (BMP15) cause increased ovulation rate and infertility in a dosage-sensitive manner. Nat Genet. 2000;25:279–83.
24. Hanrahan JP, Gregan SM, Mulsant P, Mullen M, Davis GH, Powell R, Galloway SM. Mutations in the genes for oocyte-derived growth factors GDF9 and BMP15 are associated with both increased ovulation rate and sterility in Cambridge and Belclare sheep (Ovis aries). Biol Reprod. 2004;70:900–9.
25. Davis BJ, McCurdy EA, Miller BD, Lucier GW, Tritscher AM. Ovarian tumors in rats induced by chronic 2,3,7,8-tetrachlorodibenzo-p-dioxin treatment. Cancer Res. 2000;60: 5414–9.
26. Funayama Y, Sasano H, Suzuki T, Tamura M, Fukaya T, Yajima A. Cell turnover in normal cycling human ovary. J Clin Endocrinol Metab. 1996;81:828–34.
27. Rivera OE, Varayoud J, Rodriguez HA, Munoz-de-Toro M, Luque EH. Neonatal exposure to bisphenol A or diethylstilbestrol alters the ovarian follicular dynamics in the lamb. Reprod Toxicol. 2011;32:304–12.
28. Desmeules P, Devine PJ. Characterizing the ovotoxicity of cyclophosphamide metabolites on cultured mouse ovaries. Toxicol Sci. 2006;90:500–9.
29. Ambruosi B, Uranio MF, Sardanelli AM, Pocar P, Martino NA, Paternoster MS, Amati F, Dell'Aquila ME. In vitro acute exposure to DEHP affects oocyte meiotic maturation, energy and oxidative stress parameters in a large animal model. PLoS One. 2011;6, e27452.
30. Borgeest C, Miller KP, Gupta R, Greenfeld C, Hruska KS, Hoyer P, Flaws JA. Methoxychlor-induced atresia in the mouse involves Bcl-2 family members, but not gonadotropins or estradiol. Biol Reprod. 2004;70:1828–35.
31. Lenie S, Cortvrindt R, Eichenlaub-Ritter U, Smitz J. Continuous exposure to bisphenol A during in vitro follicular development induces meiotic abnormalities. Mutat Res. 2008;651: 71–81.
32. Hu X, Christian PJ, Thompson KE, Sipes IG, Hoyer PB. Apoptosis induced in rats by 4-vinylcyclohexene diepoxide is associated with activation of the caspase cascades. Biol Reprod. 2001;65:87–93.
33. Hu X, Christian P, Sipes IG, Hoyer PB. Expression and redistribution of cellular Bad, Bax, and Bcl-X(L) protein is associated with VCD-induced ovotoxicity in rats. Biol Reprod. 2001;65:1489–95.

34. Miller KP, Gupta RK, Greenfeld CR, Babus JK, Flaws JA. Methoxychlor directly affects ovarian antral follicle growth and atresia through Bcl-2- and Bax-mediated pathways. Toxicol Sci. 2005;88:213–21.
35. Springer LN, Tilly JL, Sipes IG, Hoyer PB. Enhanced expression of bax in small preantral follicles during 4-vinylcyclohexene diepoxide-induced ovotoxicity in the rat. Toxicol Appl Pharmacol. 1996;139:402–10.
36. Gannon AM, Stampfli MR, Foster WG. Cigarette smoke exposure leads to follicle loss via an alternative ovarian cell death pathway in a mouse model. Toxicol Sci. 2012;125:274–84.
37. Stocco C. Aromatase expression in the ovary: hormonal and molecular regulation. Steroids. 2008;73:473–87.
38. Luisi S, Florio P, Reis FM, Petraglia F. Inhibins in female and male reproductive physiology: role in gametogenesis, conception, implantation and early pregnancy. Hum Reprod Update. 2005;11:123–35.
39. Cyr DG, Eales JG. Interrelationships between thyroidal and reproductive endocrine systems in fish. Rev Fish Biol Fisheries. 1996;6:165–200.
40. Brown SB, Adams BA, Cyr DG, Eales JG. Contaminant effects on the teleost fish thyroid. Environ Toxicol Chem. 2004;23:1680–701.
41. Cyr DG, Eales JG. Influence of thyroidal status on ovarian-function in Rainbow-Trout, Salmo-Gairdneri. J Exp Zool. 1988;248:81–7.
42. Cyr DG, Eales JG. Invitro effects of thyroid-hormones on gonadotropin-induced Estradiol-17-Beta secretion by ovarian follicles of Rainbow-Trout, Salmo-Gairdneri. Gen Comp Endocrinol. 1988;69:80–7.
43. Cyr DG, Eales JG. T3 enhancement of in vitro Estradiol-17-Beta secretion by oocytes of Rainbow-Trout, Salmo-Gairdneri, is dependent on cAMP. Fish Physiol Biochem. 1989;6:255–9.
44. de Montgolfier B, Audet C, Cyr DG. Regulation of the connexin 43 promoter in the brook trout testis: role of the thyroid hormones and cAMP. Gen Comp Endocrinol. 2011;170:110–8.
45. Dittrich R, Beckmann MW, Oppelt PG, Hoffmann I, Lotz L, Kuwert T, Mueller A. Thyroid hormone receptors and reproduction. J Reprod Immunol. 2011;90:58–66.
46. Reh A, Grifo J, Danoff A. What is a normal thyroid-stimulating hormone (TSH) level? Effects of stricter TSH thresholds on pregnancy outcomes after in vitro fertilization. Fertil Steril. 2010;94:2920–2.
47. Benson RC, Dailey ME. The menstrual pattern in hyperthyroidism and subsequent posttherapy hypothyroidism. Surg Gynecol Obstet. 1955;100:19–26.
48. Akande EO, Hockaday TD. Plasma oestrogen and luteinizing hormone concentrations in thyrotoxic menstrual disturbance. Proc R Soc Med. 1972;65:789–90.
49. Slebodzinski AB. Ovarian iodide uptake and triiodothyronine generation in follicular fluid. The enigma of the thyroid ovary interaction. Domest Anim Endocrinol. 2005;29:97–103.
50. Celik C, Abali R, Tasdemir N, Guzel S, Yuksel A, Aksu E, Yilmaz M. Is subclinical hypothyroidism contributing dyslipidemia and insulin resistance in women with polycystic ovary syndrome? Gynecol Endocrinol. 2012;28:615–8.
51. Unuane D, Tournaye H, Velkeniers B, Poppe K. Endocrine disorders & female infertility. Best Pract Res Clin Endocrinol Metab. 2011;25:861–73.
52. Adriance MC, Inman JL, Petersen OW, Bissell MJ. Myoepithelial cells: good fences make good neighbors. Breast Cancer Res. 2005;7:190–7.
53. Fitko R, Kucharski J, Szlezyngier B, Jana B. The concentration of GnRH in hypothalamus, LH and FSH in pituitary, LH, PRL and sex steroids in peripheral and ovarian venous plasma of hypo- and hyperthyroid, cysts-bearing gilts. Anim Reprod Sci. 1996;45:123–38.
54. Maruo T, Hayashi M, Matsuo H, Yamamoto T, Okada H, Mochizuki M. The role of thyroid hormone as a biological amplifier of the actions of follicle-stimulating hormone in the functional differentiation of cultured porcine granulosa cells. Endocrinology. 1987;121:1233–41.

55. Crofton KM. Thyroid disrupting chemicals: mechanisms and mixtures. Int J Androl. 2008; 31:209–23.
56. Gershon E, Plaks V, Dekel N. Gap junctions in the ovary: expression, localization and function. Mol Cell Endocrinol. 2008;282:18–25.
57. Kidder GM, Mhawi AA. Gap junctions and ovarian folliculogenesis. Reproduction. 2002;123: 613–20.
58. Kidder GM, Vanderhyden BC. Bidirectional communication between oocytes and follicle cells: ensuring oocyte developmental competence. Can J Physiol Pharmacol. 2010;88: 399–413.
59. Cyr DG. Connexins and pannexins: coordinating cellular communication in the testis and epididymis. Spermatogenesis. 2011;1:325–38.
60. Beyer EC, Paul DL, Goodenough DA. Connexin family of gap junction proteins. J Membr Biol. 1990;116:187–94.
61. Goodenough DA. Gap junction dynamics and intercellular communication. Pharmacol Rev. 1978;30:383–92.
62. Valdimarsson G, De Sousa PA, Kidder GM. Coexpression of gap junction proteins in the cumulus-oocyte complex. Mol Reprod Dev. 1993;36:7–15.
63. Simon AM, Goodenough DA, Li E, Paul DL. Female infertility in mice lacking connexin 37. Nature. 1997;385:525–9.
64. Veitch GI, Gittens JE, Shao Q, Laird DW, Kidder GM. Selective assembly of connexin37 into heterocellular gap junctions at the oocyte/granulosa cell interface. J Cell Sci. 2004;117:2699–707.
65. Carabatsos MJ, Sellitto C, Goodenough DA, Albertini DF. Oocyte-granulosa cell heterologous gap junctions are required for the coordination of nuclear and cytoplasmic meiotic competence. Dev Biol. 2000;226:167–79.
66. Edry I, Sela-Abramovich S, Dekel N. Meiotic arrest of oocytes depends on cell-to-cell communication in the ovarian follicle. Mol Cell Endocrinol. 2006;252:102–6.
67. Sela-Abramovich S, Edry I, Galiani D, Nevo N, Dekel N. Disruption of gap junctional communication within the ovarian follicle induces oocyte maturation. Endocrinology. 2006; 147:2280–6.
68. Granot I, Dekel N. The ovarian gap junction protein connexin43: regulation by gonadotropins. Trends Endocrinol Metab. 2002;13:310–3.
69. Wiesen JF, Midgley Jr AR. Expression of connexin 43 gap junction messenger ribonucleic acid and protein during follicular atresia. Biol Reprod. 1994;50:336–48.
70. Granot I, Dekel N. Developmental expression and regulation of the gap junction protein and transcript in rat ovaries. Mol Reprod Dev. 1997;47:231–9.
71. Granot I, Dekel N. Phosphorylation and expression of connexin-43 ovarian gap junction protein are regulated by luteinizing hormone. J Biol Chem. 1994;269:30502–9.
72. Kalma Y, Granot I, Galiani D, Barash A, Dekel N. Luteinizing hormone-induced connexin 43 down-regulation: inhibition of translation. Endocrinology. 2004;145:1617–24.
73. Reaume AG, de Sousa PA, Kulkarni S, Langille BL, Zhu D, Davies TC, Juneja SC, Kidder GM, Rossant J. Cardiac malformation in neonatal mice lacking connexin43. Science. 1995; 267:1831–4.
74. Juneja SC, Barr KJ, Enders GC, Kidder GM. Defects in the germ line and gonads of mice lacking connexin43. Biol Reprod. 1999;60:1263–70.
75. Flenniken AM, Osborne LR, Anderson N, Ciliberti N, Fleming C, Gittens JE, Gong XQ, Kelsey LB, Lounsbury C, Moreno L, Nieman BJ, Peterson K, Qu D, Roscoe W, Shao Q, Tong D, Veitch GI, Voronina I, Vukobradovic I, Wood GA, Zhu Y, Zirngibl RA, Aubin JE, Bai D, Bruneau BG, Grynpas M, Henderson JE, Henkelman RM, McKerlie C, Sled JG, Stanford WL, Laird DW, Kidder GM, Adamson SL, Rossant J. A Gja1 missense mutation in a mouse model of oculodentodigital dysplasia. Development. 2005;132:4375–86.
76. Tong D, Colley D, Thoo R, Li TY, Plante I, Laird DW, Bai D, Kidder GM. Oogenesis defects in a mutant mouse model of oculodentodigital dysplasia. Dis Model Mech. 2009;2:157–67.

77. Sasson R, Amsterdam A. Pleiotropic anti-apoptotic activity of glucocorticoids in ovarian follicular cells. Biochem Pharmacol. 2003;66:1393–401.
78. Sasson R, Shinder V, Dantes A, Land A, Amsterdam A. Activation of multiple signal transduction pathways by glucocorticoids: protection of ovarian follicular cells against apoptosis. Biochem Biophys Res Commun. 2003;311:1047–56.
79. Johnson ML, Redmer DA, Reynolds LP, Grazul-Bilska AT. Expression of gap junctional proteins connexin 43, 32, and 26 throughout follicular development and atresia in cows. Endocrine. 1999;10:43–51.
80. Krysko DV, Mussche S, Leybaert L, D'Herde K. Gap junctional communication and connexin43 expression in relation to apoptotic cell death and survival of granulosa cells. J Histochem Cytochem. 2004;52:1199–207.
81. Kishta O, Adeeko A, Li D, Luu T, Brawer JR, Morales C, Hermo L, Robaire B, Hales BF, Barthelemy J, Cyr DG, Trasler JM. In utero exposure to tributyltin chloride differentially alters male and female fetal gonad morphology and gene expression profiles in the Sprague–Dawley rat. Reprod Toxicol. 2007;23:1–11.
82. Ke FC, Fang SH, Lee MT, Sheu SY, Lai SY, Chen YJ, Huang FL, Wang PS, Stocco DM, Hwang JJ. Lindane, a gap junction blocker, suppresses FSH and transforming growth factor beta1-induced connexin43 gap junction formation and steroidogenesis in rat granulosa cells. J Endocrinol. 2005;184:555–66.
83. Devine PJ, Rajapaksa KS, Hoyer PB. In vitro ovarian tissue and organ culture: a review. Front Biosci. 2002;7:d1979–1989.
84. Holt JE, Jackson A, Roman SD, Aitken RJ, Koopman P, McLaughlin EA. CXCR4/SDF1 interaction inhibits the primordial to primary follicle transition in the neonatal mouse ovary. Dev Biol. 2006;293:449–60.
85. Igawa Y, Keating AF, Rajapaksa KS, Sipes IG, Hoyer PB. Evaluation of ovotoxicity induced by 7, 12-dimethylbenz[a]anthracene and its 3,4-diol metabolite utilizing a rat in vitro ovarian culture system. Toxicol Appl Pharmacol. 2009;234:361–9.
86. Kezele P, Nilsson EE, Skinner MK. Keratinocyte growth factor acts as a mesenchymal factor that promotes ovarian primordial to primary follicle transition. Biol Reprod. 2005; 73:967–73.
87. Devine PJ, Hoyer PB, Keating A. Current methods in investigating the development of the female reproductive system. In: Vaillancourt C, Lafond J, editors. Human embryogenesis: methods and protocols. Totawa, NH: Humana Press; 2009. p. 137–57.
88. Devine PJ, Petrillo SK, Cortvrindt R. In vitro ovarian model systems to study toxicology. In: Hoyer PB, editor. Female reproductive toxicology. 2nd ed. London: Elsevier; 2010. p. 543–59.
89. Durlinger AL, Gruijters MJ, Kramer P, Karels B, Ingraham HA, Nachtigal MW, Uilenbroek JT, Grootegoed JA, Themmen AP. Anti-Mullerian hormone inhibits initiation of primordial follicle growth in the mouse ovary. Endocrinology. 2002;143:1076–84.
90. Petrillo SK, Desmeules P, Truong TQ, Devine PJ. Detection of DNA damage in oocytes of small ovarian follicles following phosphoramide mustard exposures of cultured rodent ovaries in vitro. Toxicol Appl Pharmacol. 2011;253:94–102.
91. Colvin OM. An overview of cyclophosphamide development and clinical applications. Curr Pharm Des. 1999;5:555–60.
92. Plowchalk DR, Mattison DR. Phosphoramide mustard is responsible for the ovarian toxicity of cyclophosphamide. Toxicol Appl Pharmacol. 1991;107:472–81.
93. Rogakou EP, Pilch DR, Orr AH, Ivanova VS, Bonner WM. DNA double-stranded breaks induce histone H2AX phosphorylation on serine 139. J Biol Chem. 1998;273:5858–68.
94. Ataya KM, Valeriote FA, Ramahi-Ataya AJ. Effect of cyclophosphamide on the immature rat ovary. Cancer Res. 1989;49:1660–4.
95. Borman SM, VanDePol BJ, Kao S, Thompson KE, Sipes IG, Hoyer PB. A single dose of the ovotoxicant 4-vinylcyclohexene diepoxide is protective in rat primary ovarian follicles. Toxicol Appl Pharmacol. 1999;158:244–52.

96. Fernandez SM, Keating AF, Christian PJ, Sen N, Hoying JB, Brooks HL, Hoyer PB. Involvement of the KIT/KITL signaling pathway in 4-vinylcyclohexene diepoxide-induced ovarian follicle loss in rats. Biol Reprod. 2008;79:318–27.
97. Keating AF, Mark CJ, Sen N, Sipes IG, Hoyer PB. Effect of phosphatidylinositol-3 kinase inhibition on ovotoxicity caused by 4-vinylcyclohexene diepoxide and 7, 12-dimethylbenz[a] anthracene in neonatal rat ovaries. Toxicol Appl Pharmacol. 2009;241:127–34.
98. Keating AF, Fernandez SM, Mark-Kappeler CJ, Sen N, Sipes IG, Hoyer PB. Inhibition of PIK3 signaling pathway members by the ovotoxicant 4-vinylcyclohexene diepoxide in rats. Biol Reprod. 2011;84:743–51.
99. Gupta RK, Meachum S, Hernandez-Ochoa I, Peretz J, Yao HH, Flaws JA. Methoxychlor inhibits growth of antral follicles by altering cell cycle regulators. Toxicol Appl Pharmacol. 2009;240:1–7.
100. Muskhelishvili L, Wingard SK, Latendresse JR. Proliferating cell nuclear antigen—a marker for ovarian follicle counts. Toxicol Pathol. 2005;33:365–8.
101. Picut CA, Swanson CL, Scully KL, Roseman VC, Parker RF, Remick AK. Ovarian follicle counts using proliferating cell nuclear antigen (PCNA) and semi-automated image analysis in rats. Toxicol Pathol. 2008;36:674–9.
102. Muskhelishvili L, Freeman LD, Latendresse JR, Bucci TJ. An immunohistochemical label to facilitate counting of ovarian follicles. Toxicol Pathol. 2002;30:400–2.
103. East IJ, Gulyas BJ, Dean J. Monoclonal antibodies to the murine zona pellucida protein with sperm receptor activity: effects on fertilization and early development. Dev Biol. 1985; 109:268–73.
104. Yeh J, Kim B, Liang YJ, Peresie J. Mullerian inhibiting substance as a novel biomarker of cisplatin-induced ovarian damage. Biochem Biophys Res Commun. 2006;348:337–44.
105. Gupta RK, Miller KP, Babus JK, Flaws JA. Methoxychlor inhibits growth and induces atresia of antral follicles through an oxidative stress pathway. Toxicol Sci. 2006;93:382–9.
106. Plante I, Charbonneau M, Cyr DG. Decreased gap junctional intercellular communication in hexachlorobenzene-induced gender-specific hepatic tumor formation in the rat. Carcinogenesis. 2002;23:1243–9.
107. Toler CR, Taylor DD, Gercel-Taylor C. Loss of communication in ovarian cancer. Am J Obstet Gynecol. 2006;194:e27–31.
108. Lucke C, Siebert S, Mayr D, Mayerhofer A. Connexin expression by human granulosa cell tumors: identification of connexin 32 as a tumor signature. Cancer Biomark. 2010; 8:137–44.
109. Li J, Wood 3rd WH, Becker KG, Weeraratna AT, Morin PJ. Gene expression response to cisplatin treatment in drug-sensitive and drug-resistant ovarian cancer cells. Oncogene. 2007;26:2860–72.
110. Chen X, Aravindakshan J, Yang Y, Sairam MR. Early alterations in ovarian surface epithelial cells and induction of ovarian epithelial tumors triggered by loss of FSH receptor. Neoplasia. 2007;9:521–31.
111. Aravindakshan J, Chen X, Sairam MR. Differential expression of claudin family proteins in mouse ovarian serous papillary epithelial adenoma in aging FSH receptor-deficient mutants. Neoplasia. 2006;8:984–94.
112. Fisher JS. Environmental anti-androgens and male reproductive health: focus on phthalates and testicular dysgenesis syndrome. Reproduction. 2004;127:305–15.
113. Diamanti-Kandarakis E, Bourguignon JP, Giudice LC, Hauser R, Prins GS, Soto AM, Zoeller RT, Gore AC. Endocrine-disrupting chemicals: an Endocrine Society scientific statement. Endocr Rev. 2009;30:293–342.
114. Aksglaede L, Sorensen K, Petersen JH, Skakkebaek NE, Juul A. Recent decline in age at breast development: the Copenhagen Puberty Study. Pediatrics. 2009;123:e932–939.
115. Castellino N, Bellone S, Rapa A, Vercellotti A, Binotti M, Petri A, Bona G. Puberty onset in Northern Italy: a random sample of 3597 Italian children. J Endocrinol Invest. 2005; 28:589–94.

116. Euling SY, Herman-Giddens ME, Lee PA, Selevan SG, Juul A, Sorensen TI, Dunkel L, Himes JH, Teilmann G, Swan SH. Examination of US puberty-timing data from 1940 to 1994 for secular trends: panel findings. Pediatrics. 2008;121 Suppl 3:S172–191.

117. Mul D, Fredriks AM, van Buuren S, Oostdijk W, Verloove-Vanhorick SP, Wit JM. Pubertal development in The Netherlands 1965–1997. Pediatr Res. 2001;50:479–86.

118. Bocchinfuso WP, Lindzey JK, Hewitt SC, Clark JA, Myers PH, Cooper R, Korach KS. Induction of mammary gland development in estrogen receptor-alpha knockout mice. Endocrinology. 2000;141:2982–94.

119. Sternlicht MD. Key stages in mammary gland development: the cues that regulate ductal branching morphogenesis. Breast Cancer Res. 2006;8:201.

120. Mouritsen A, Aksglaede L, Sorensen K, Mogensen SS, Leffers H, Main KM, Frederiksen H, Andersson AM, Skakkebaek NE, Juul A. Hypothesis: exposure to endocrine-disrupting chemicals may interfere with timing of puberty. Int J Androl. 2010;33:346–59.

121. Chumlea WC, Schubert CM, Roche AF, Kulin HE, Lee PA, Himes JH, Sun SS. Age at menarche and racial comparisons in US girls. Pediatrics. 2003;111:110–3.

122. Herman-Giddens ME, Slora EJ, Wasserman RC, Bourdony CJ, Bhapkar MV, Koch GG, Hasemeier CM. Secondary sexual characteristics and menses in young girls seen in office practice: a study from the Pediatric Research in Office Settings network. Pediatrics. 1997; 99:505–12.

123. Crain DA, Janssen SJ, Edwards TM, Heindel J, Ho SM, Hunt P, Iguchi T, Juul A, McLachlan JA, Schwartz J, Skakkebaek N, Soto AM, Swan S, Walker C, Woodruff TK, Woodruff TJ, Giudice LC, Guillette Jr LJ. Female reproductive disorders: the roles of endocrine-disrupting compounds and developmental timing. Fertil Steril. 2008;90:911–40.

124. Gray J, Evans N, Taylor B, Rizzo J, Walker M. State of the evidence: the connection between breast cancer and the environment. Int J Occup Environ Health. 2009;15:43–78.

125. Ghajar CM, Bissell MJ. Extracellular matrix control of mammary gland morphogenesis and tumorigenesis: insights from imaging. Histochem Cell Biol. 2008;130:1105–18.

126. Neville MC, Medina D, Monks J, Hovey RC. The mammary fat pad. J Mammary Gland Biol Neoplasia. 1998;3:109 16.

127. Hennighausen L, Robinson GW. Information networks in the mammary gland. Nat Rev Mol Cell Biol. 2005;6:715–25.

128. Timpl R. Macromolecular organization of basement membranes. Curr Opin Cell Biol. 1996;8:618–24.

129. Hens JR, Wysolmerski JJ. Key stages of mammary gland development: molecular mechanisms involved in the formation of the embryonic mammary gland. Breast Cancer Res. 2005; 7:220–4.

130. Robinson GW. Cooperation of signalling pathways in embryonic mammary gland development. Nat Rev Genet. 2007;8:963–72.

131. Hinck L, Silberstein GB. Key stages in mammary gland development: the mammary end bud as a motile organ. Breast Cancer Res. 2005;7:245–51.

132. Daniel CW, Silberstein GB, Strickland P. Direct action of 17 beta-estradiol on mouse mammary ducts analyzed by sustained release implants and steroid autoradiography. Cancer Res. 1987;47:6052–7.

133. Korach KS, Couse JF, Curtis SW, Washburn TF, Lindzey J, Kimbro KS, Eddy EM, Migliaccio S, Snedeker SM, Lubahn DB, Schomberg DW, Smith EP. Estrogen receptor gene disruption: molecular characterization and experimental and clinical phenotypes. Recent Prog Horm Res. 1996;51:159–86.

134. Mallepell S, Krust A, Chambon P, Brisken C. Paracrine signaling through the epithelial estrogen receptor alpha is required for proliferation and morphogenesis in the mammary gland. Proc Natl Acad Sci U S A. 2006;103:2196–201.

135. Aupperlee MD, Drolet AA, Durairaj S, Wang W, Schwartz RC, Haslam SZ. Strain-specific differences in the mechanisms of progesterone regulation of murine mammary gland development. Endocrinology. 2009;150:1485–94.

136. Neville MC, McFadden TB, Forsyth I. Hormonal regulation of mammary differentiation and milk secretion. J Mammary Gland Biol Neoplasia. 2002;7:49–66.
137. McManaman JL, Neville MC. Mammary physiology and milk secretion. Adv Drug Deliv Rev. 2003;55:629–41.
138. Reversi A, Cassoni P, Chini B. Oxytocin receptor signaling in myoepithelial and cancer cells. J Mammary Gland Biol Neoplasia. 2005;10:221–9.
139. Stein T, Salomonis N, Gusterson BA. Mammary gland involution as a multi-step process. J Mammary Gland Biol Neoplasia. 2007;12:25–35.
140. Vandenberg LN, Maffini MV, Wadia PR, Sonnenschein C, Rubin BS, Soto AM. Exposure to environmentally relevant doses of the xenoestrogen bisphenol-A alters development of the fetal mouse mammary gland. Endocrinology. 2007;148:116–27.
141. El Sheikh Saad H, Meduri G, Phrakonkham P, Berges R, Vacher S, Djallali M, Auger J, Canivenc-Lavier MC, Perrot-Applanat M. Abnormal peripubertal development of the rat mammary gland following exposure in utero and during lactation to a mixture of genistein and the food contaminant vinclozolin. Reprod Toxicol. 2011;32:15–25.
142. Enoch RR, Stanko JP, Greiner SN, Youngblood GL, Rayner JL, Fenton SE. Mammary gland development as a sensitive end point after acute prenatal exposure to an atrazine metabolite mixture in female Long-Evans rats. Environ Health Perspect. 2007;115:541–7.
143. Markey CM, Luque EH, Munoz De Toro M, Sonnenschein C, Soto AM. In utero exposure to bisphenol A alters the development and tissue organization of the mouse mammary gland. Biol Reprod. 2001;65:1215–23.
144. Moral R, Santucci-Pereira J, Wang R, Russo IH, Lamartiniere CA, Russo J. In utero exposure to butyl benzyl phthalate induces modifications in the morphology and the gene expression profile of the mammary gland: an experimental study in rats. Environ Health. 2011;10:5.
145. Rayner JL, Enoch RR, Fenton SE. Adverse effects of prenatal exposure to atrazine during a critical period of mammary gland growth. Toxicol Sci. 2005;87:255–66.
146. Ayyanan A, Laribi O, Schuepbach-Mallepell S, Schrick C, Gutierrez M, Tanos T, Lefebvre G, Rougemont J, Yalcin-Ozuysal O, Brisken C. Perinatal exposure to bisphenol a increases adult mammary gland progesterone response and cell number. Mol Endocrinol. 2011; 25:1915–23.
147. Vandenberg LN, Maffini MV, Schaeberle CM, Ucci AA, Sonnenschein C, Rubin BS, Soto AM. Perinatal exposure to the xenoestrogen bisphenol-A induces mammary intraductal hyperplasias in adult CD-1 mice. Reprod Toxicol. 2008;26:210–9.
148. Richter CA, Birnbaum LS, Farabollini F, Newbold RR, Rubin BS, Talsness CE, Vandenbergh JG, Walser-Kuntz DR, vom Saal FS. In vivo effects of bisphenol A in laboratory rodent studies. Reprod Toxicol. 2007;24:199–224.
149. Brown NM, Lamartiniere CA. Xenoestrogens alter mammary gland differentiation and cell proliferation in the rat. Environ Health Perspect. 1995;103:708–13.
150. Colerangle JB, Roy D. Profound effects of the weak environmental estrogen-like chemical bisphenol A on the growth of the mammary gland of Noble rats. J Steroid Biochem Mol Biol. 1997;60:153–60.
151. Vorderstrasse BA, Fenton SE, Bohn AA, Cundiff JA, Lawrence BP. A novel effect of dioxin: exposure during pregnancy severely impairs mammary gland differentiation. Toxicol Sci. 2004;78:248–57.
152. Palmer CA, Neville MC, Anderson SM, McManaman JL. Analysis of lactation defects in transgenic mice. J Mammary Gland Biol Neoplasia. 2006;11:269–82.
153. Koch HM, Preuss R, Drexler H, Angerer J. Exposure of nursery school children and their parents and teachers to di-n-butylphthalate and butylbenzylphthalate. Int Arch Occup Environ Health. 2005;78:223–9.
154. Meeker JD, Sathyanarayana S, Swan SH. Phthalates and other additives in plastics: human exposure and associated health outcomes. Philos Trans R Soc Lond B Biol Sci. 2009;364:2097–113.
155. Schell LM, Gallo MV. Relationships of putative endocrine disruptors to human sexual maturation and thyroid activity in youth. Physiol Behav. 2010;99:246–53.

156. Silva MJ, Barr DB, Reidy JA, Malek NA, Hodge CC, Caudill SP, Brock JW, Needham LL, Calafat AM. Urinary levels of seven phthalate metabolites in the U.S. population from the National Health and Nutrition Examination Survey (NHANES) 1999–2000. Environ Health Perspect. 2004;112:331–8.
157. Grill CJ, Cohick WS, Sherman AR. Postpubertal development of the rat mammary gland is preserved during iron deficiency. J Nutr. 2001;131:1444–8.
158. Plante I, Laird DW. Decreased levels of connexin43 result in impaired development of the mammary gland in a mouse model of oculodentodigital dysplasia. Dev Biol. 2008;318:312–22.
159. Plante I, Stewart MK, Laird DW. Evaluation of mammary gland development and function in mouse models. J Vis Exp. 2011; (53).
160. Silberstein GB, Flanders KC, Roberts AB, Daniel CW. Regulation of mammary morphogenesis: evidence for extracellular matrix-mediated inhibition of ductal budding by transforming growth factor-beta 1. Dev Biol. 1992;152:354–62.
161. You L, Sar M, Bartolucci EJ, McIntyre BS, Sriperumbudur R. Modulation of mammary gland development in prepubertal male rats exposed to genistein and methoxychlor. Toxicol Sci. 2002;66:216–25.
162. Cunha GR. Role of mesenchymal-epithelial interactions in normal and abnormal development of the mammary gland and prostate. Cancer. 1994;74:1030–44.
163. Cunha GR, Hom YK. Role of mesenchymal-epithelial interactions in mammary gland development. J Mammary Gland Biol Neoplasia. 1996;1:21–35.
164. Cunha GR, Young P, Hom YK, Cooke PS, Taylor JA, Lubahn DB. Elucidation of a role for stromal steroid hormone receptors in mammary gland growth and development using tissue recombinants. J Mammary Gland Biol Neoplasia. 1997;2:393–402.
165. Daniel CW, De Ome KB, Young JT, Blair PB, Faulkin Jr LJ. The in vivo life span of normal and preneoplastic mouse mammary glands: a serial transplantation study. Proc Natl Acad Sci USA. 1968;61:53–60.
166. Naylor MJ, Ormandy CJ. Mouse strain-specific patterns of mammary epithelial ductal side branching are elicited by stromal factors. Dev Dyn. 2002;225:100–5.
167. Wiesen JF, Young P, Werb Z, Cunha GR. Signaling through the stromal epidermal growth factor receptor is necessary for mammary ductal development. Development. 1999; 126:335–44.
168. Carraway KL, Ramsauer VP, Carraway CA. Glycoprotein contributions to mammary gland and mammary tumor structure and function: roles of adherens junctions, ErbBs and membrane MUCs. J Cell Biochem. 2005;96:914–26.
169. El-Sabban ME, Abi-Mosleh LF, Talhouk RS. Developmental regulation of gap junctions and their role in mammary epithelial cell differentiation. J Mammary Gland Biol Neoplasia. 2003;8:463–73.
170. El-Saghir JA, El-Habre ET, El-Sabban ME, Talhouk RS. Connexins: a junctional crossroad to breast cancer. Int J Dev Biol. 2011;55:773–80.
171. Pitelka DR, Hamamoto ST, Duafala JG, Nemanic MK. Cell contacts in the mouse mammary gland. I Normal gland in postnatal development and the secretory cycle. J Cell Biol. 1973;56:797–818.
172. Laird DW. Life cycle of connexins in health and disease. Biochem J. 2006;394:527–43.
173. Niessen CM, Gottardi CJ. Molecular components of the adherens junction. Biochim Biophys Acta. 2008;1778:562–71.
174. Niessen CM. Tight junctions/adherens junctions: basic structure and function. J Invest Dermatol. 2007;127:2525–32.
175. Guillemot L, Paschoud S, Pulimeno P, Foglia A, Citi S. The cytoplasmic plaque of tight junctions: a scaffolding and signalling center. Biochim Biophys Acta. 2008;1778:601–13.
176. Delmas V, Pla P, Feracci H, Thiery JP, Kemler R, Larue L. Expression of the cytoplasmic domain of E-cadherin induces precocious mammary epithelial alveolar formation and affects cell polarity and cell-matrix integrity. Dev Biol. 1999;216:491–506.

177. Hatsell S, Rowlands T, Hiremath M, Cowin P. Beta-catenin and Tcfs in mammary development and cancer. J Mammary Gland Biol Neoplasia. 2003;8:145–58.
178. Radice GL, Ferreira-Cornwell MC, Robinson SD, Rayburn H, Chodosh LA, Takeichi M, Hynes RO. Precocious mammary gland development in P-cadherin-deficient mice. J Cell Biol. 1997;139:1025–32.
179. Aravindakshan J, Cyr DG. Nonylphenol alters connexin 43 levels and connexin 43 phosphorylation via an inhibition of the p38-mitogen-activated protein kinase pathway. Biol Reprod. 2005;72:1232–40.
180. Gillum N, Karabekian Z, Swift LM, Brown RP, Kay MW, Sarvazyan N. Clinically relevant concentrations of di (2-ethylhexyl) phthalate (DEHP) uncouple cardiac syncytium. Toxicol Appl Pharmacol. 2009;236:25–38.
181. Plante I, Charbonneau M, Cyr DG. Activation of the integrin-linked kinase pathway down-regulates hepatic connexin32 via nuclear Akt. Carcinogenesis. 2006;27:1923–9.
182. Plante I, Cyr DG, Charbonneau M. Involvement of the integrin-linked kinase pathway in hexachlorobenzene-induced gender-specific rat hepatocarcinogenesis. Toxicol Sci. 2005; 88:346–57.
183. Plante I, Cyr DG, Charbonneau M. Sexual dimorphism in the regulation of liver connexin32 transcription in hexachlorobenzene-treated rats. Toxicol Sci. 2007;96:47–57.
184. Sobarzo CM, Lustig L, Ponzio R, Denduchis B. Effect of di-(2-ethylhexyl) phthalate on N-cadherin and catenin protein expression in rat testis. Reprod Toxicol. 2006;22:77–86.
185. Sobarzo CM, Lustig L, Ponzio R, Suescun MO, Denduchis B. Effects of di(2-ethylhexyl) phthalate on gap and tight junction protein expression in the testis of prepubertal rats. Microsc Res Tech. 2009;72:868–77.
186. Blanchard AA, Watson PH, Shiu RP, Leygue E, Nistor A, Wong P, Myal Y. Differential expression of claudin 1, 3, and 4 during normal mammary gland development in the mouse. DNA Cell Biol. 2006;25:79–86.
187. Daniel CW, Strickland P, Friedmann Y. Expression and functional role of E- and P-cadherins in mouse mammary ductal morphogenesis and growth. Dev Biol. 1995;169:511–9.
188. Nguyen DA, Neville MC. Tight junction regulation in the mammary gland. J Mammary Gland Biol Neoplasia. 1998;3:233–46.
189. Stingl J, Raouf A, Emerman JT, Eaves CJ. Epithelial progenitors in the normal human mammary gland. J Mammary Gland Biol Neoplasia. 2005;10:49–59.
190. Monaghan P, Perusinghe N, Carlile G, Evans WH. Rapid modulation of gap junction expression in mouse mammary gland during pregnancy, lactation, and involution. J Histochem Cytochem. 1994;42:931–8.
191. Plante I, Wallis A, Shao Q, Laird DW. Milk secretion and ejection are impaired in the mammary gland of mice harboring a Cx43 mutant while expression and localization of tight and adherens junction proteins remain unchanged. Biol Reprod. 2010;82:837–47.
192. Talhouk RS, Elble RC, Bassam R, Daher M, Sfeir A, Mosleh LA, El-Khoury H, Hamoui S, Pauli BU, El-Sabban ME. Developmental expression patterns and regulation of connexins in the mouse mammary gland: expression of connexin30 in lactogenesis. Cell Tissue Res. 2005; 319:49–59.
193. Nguyen DA, Parlow AF, Neville MC. Hormonal regulation of tight junction closure in the mouse mammary epithelium during the transition from pregnancy to lactation. J Endocrinol. 2001;170:347–56.
194. Dube E, Dufresne J, Chan PT, Hermo L, Cyr DG. Assessing the role of claudins in maintaining the integrity of epididymal tight junctions using novel human epididymal cell lines. Biol Reprod. 2010;82:1119–28.
195. Richert MM, Schwertfeger KL, Ryder JW, Anderson SM. An atlas of mouse mammary gland development. J Mammary Gland Biol Neoplasia. 2000;5:227–41.
196. Schwertfeger KL, McManaman JL, Palmer CA, Neville MC, Anderson SM. Expression of constitutively activated Akt in the mammary gland leads to excess lipid synthesis during pregnancy and lactation. J Lipid Res. 2003;44:1100–12.

197. Shillingford JM, Miyoshi K, Robinson GW, Bierie B, Cao Y, Karin M, Hennighausen L. Proteotyping of mammary tissue from transgenic and gene knockout mice with immunohistochemical markers: a tool to define developmental lesions. J Histochem Cytochem. 2003;51:555–65.
198. Dickson SR, Warburton MJ. Enhanced synthesis of gelatinase and stromelysin by myoepithelial cells during involution of the rat mammary gland. J Histochem Cytochem. 1992;40:697–703.
199. Gudjonsson T, Adriance MC, Sternlicht MD, Petersen OW, Bissell MJ. Myoepithelial cells: their origin and function in breast morphogenesis and neoplasia. J Mammary Gland Biol Neoplasia. 2005;10:261–72.
200. Warburton MJ, Mitchell D, Ormerod EJ, Rudland P. Distribution of myoepithelial cells and basement membrane proteins in the resting, pregnant, lactating, and involuting rat mammary gland. J Histochem Cytochem. 1982;30:667–76.
201. Krause S, Maffini MV, Soto AM, Sonnenschein C. A novel 3D in vitro culture model to study stromal-epithelial interactions in the mammary gland. Tissue Eng Part C Methods. 2008;14:261–71.
202. Swamydas M, Eddy JM, Burg KJ, Dreau D. Matrix compositions and the development of breast acini and ducts in 3D cultures. In Vitro Cell Dev Biol Anim. 2010;46:673–84.
203. Wang X, Sun L, Maffini MV, Soto A, Sonnenschein C, Kaplan DL. A complex 3D human tissue culture system based on mammary stromal cells and silk scaffolds for modeling breast morphogenesis and function. Biomaterials. 2010;31:3920–9.

Chapter 8
Immunohistochemistry on Rodent Circulatory System: Its Possible Use in Investigating Hypertension

Chun-Yi Ng, Yusof Kamisah, and Kamsiah Jaarin

Introduction

Vascular wall inflammation is a cardinal component in atherogenesis, from the lesion initiation to progression to the ultimate plaque rupture. Nevertheless, inflammation is also found in other diseases associated with vascular dysfunction including hypertension. The key role of vascular wall inflammation in the pathogenesis of hypertension has been increasingly recognised recently [1–3]. Inflammation may impair function of vascular wall in blood pressure regulation. Many studies have suggested the involvement of inflammation in the initiation as well as progression of hypertension. More than 28 years ago, Khraibi and colleagues [4] have found that chronic immunosuppressive therapy attenuated blood pressure elevation in Okamoto spontaneously hypertensive rats, suggesting the disease might be due to inflammation. In the following year, Norman and co-workers [5] demonstrated the important role of immunological dysfunction in spontaneous hypertension. Dzielak [6] has identified alterations in serum immunoglobulin levels and immune dysfunction alterations in both patients with hypertension and models of spontaneous hypertension. Besides, the interaction between monocytes and endothelial cells was observed to be increased in hypertensive patients [7]. Kampus et al. [8] have also reported an increase in C-reactive proteins and vascular wall stiffness in untreated hypertensive patients. Recently, Elmarakby and colleagues [9] showed that the progression of hypertension in spontaneously hypertensive rats was slowed down by attenuating inflammation. The precise mechanism(s) by which inflammation causes hypertension is however still uncertain. Rodent's circulatory system hence represents an important model for studying the role of inflammation in raising blood pressure. In this chapter, we

C.-Y. Ng • Y. Kamisah • K. Jaarin (✉)
Faculty of Medicine, Department of Pharmacology, The National University of Malaysia,
50300 Kuala Lumpur, Malaysia
e-mail: kamsiah@medic.ukm.my

© Springer Science+Business Media New York 2016   147
S.A. Aziz, R. Mehta (eds.), *Technical Aspects of Toxicological
Immunohistochemistry*, DOI 10.1007/978-1-4939-1516-3_8

discuss the inflammatory markers that are useful to investigate the role of inflammation in hypertension in rodent's circulatory system by immunohistochemistry.

Important Vascular Inflammatory Markers

Markers should be chosen based on their possible association between inflammation and the disease of interest. Inappropriate choice of markers represents a major pitfall in the use of immunohistochemistry as it may result in pointless interpretations.

Vascular Cell Adhesion Molecule-1

Vascular cell adhesion molecule-1 (VCAM-1) was first suspected as an inducible endothelial cell surface glycoprotein that appeared to mediate adhesion of melanoma cells (Rice and Bevilacqua 1989). This undescribed adhesion molecule was then identified to belong to the immunoglobulin gene superfamily using cloning approach (Osborn et al. 1989). In the following year, Carlos et al. (1990) demonstrated the adhesion of peripheral blood lymphocyte to cytokine-stimulated human umbilical vein endothelial cell (HUVEC) was dependent on VCAM-1, proposing that VCAM-1 may be an important component of lymphocyte emigration at sites of inflammation or immune reaction in vivo. These findings suggested that the expression of VCAM-1 is inducible, and treatment of endothelial cells with inflammatory cytokines such as interleukin-1 (IL-1), tumour necrosis factor-alpha (TNF-α), or bacterial lipopolysaccharide (LPS) enhanced its expression.

VCAM-1 is a type 1 transmembrane glycoprotein and a member of the immunoglobulin superfamily. Since it is categorised in this family, VCAM-1 contains immunoglobulin-like domains in its structure. These immunoglobulin-like domains are expressed on the extracellular region of the structure which is anchored to the surface of endothelial cell by a generally short cytoplasmic tail. There are two different forms of VCAM-1 in human and rodents. In human, VCAM-1 contains either seven immunoglobulin-like domains or a protein with only domain 1–3 and 5–7 (Osborn et al. 1992). Domains 1 and 4 are the regions where its ligands bind to (Chuluyan et al. 1991). So, the seven-domain VCAM-1 has two ligand binding sites. However in rodents, VCAM-1 only exists as a seven-domain transmembrane protein. The primary ligand of VCAM-1 is $\alpha4\beta1$ integrin (Very Late Antigen-4, VLA-4) which is constitutively expressed on lymphocytes, monocytes, and eosinophils (Osborn et al. 1992). It is expressed by lymph node endothelium and is induced on endothelium in inflammatory sites. VCAM-1 expression is low or absent on endothelial cells under basal conditions. Its expression is rapidly induced by pro-inflammatory mediators (Carlos et al. 1990; [10,11]).

VCAM-1 is involved in both normal physiological processes and disease pathogenesis. VCAM-1 functions as one of the endothelial cell adhesion molecules that mediate leukocytes binding. Interestingly, VCAM-1 can mediate both initial tethering adhesion and firm adhesion of circulating leukocytes on cell surface, depending on the avidity status of $\alpha4\beta1$ integrin [12]. Therefore, VCAM-1 is considered as an important adhesion system for regulating leukocytes migration. Increased expression of VCAM-1 on endothelial cells is a common process in response to inflammation. The expression of adhesion molecules (including both VCAM-1 and intercellular adhesion molecule-1, ICAM-1) has been well documented in atherosclerotic lesions [13–15]. Attachment of mononuclear leukocytes, i.e. macrophages and T-lymphocytes, to endothelial lining is the first steps in atherogenesis, leading to inflammatory responses in the vascular wall. This accumulation of these cells is regulated by cell adhesion molecules, including VCAM-1 and ICAM-1. Attachment of monocytes and lymphocytes to endothelial cells initiates the process of atherosclerosis. After transmigration into the vascular wall, leukocytes release various bioactive mediators to facilitate the accumulation of fatty streaks and formation of atherosclerotic plaque as well as proliferation of smooth muscle cells. Therefore, inhibition of adhesion molecules can delay the development of atherosclerosis, suggesting a perspective way for therapeutic intervention to treat cardiovascular diseases [16].

Aside from atherosclerosis, the involvement of adhesion molecules in the development of hypertension has also been increasingly paid attention to (Parissis et al. 2001; [17–19]). In fact, inflammatory VCAM-1 may act as a link between hypertension and an increased risk for atherosclerosis [20]. Earlier studies have demonstrated that angiotensin II, the increased circulatory level of which is associated with hypertension, provokes endothelial activation and VCAM-1 expression probably via the pro-inflammatory nuclear factor-kappa B (NF-κB) signalling pathway [21,22]. In our recent experiment, immunohistochemical staining of the endothelial linings of the aorta showed an intense positive staining for VCAM-1 in hypertensive rodents compared to normotensive rodents (Fig. 8.1). There was also a significant positive correlation between blood pressure and VCAM-1 density on endothelial cells in hypertensive rats. The result suggests that augmented VCAM-1 expression may reflect inflammation and endothelial dysfunction and hence impairing the regulation of blood pressure [23]. Its expression was also elevated in hyperhomocysteinemic rats, which was positively correlated to oxidative stress and plasma homocysteine level [24].

Intercellular Adhesion Molecule-1

Intercellular cell adhesion molecule-1 (ICAM-1) is a cell surface glycoprotein expressed on multiple cell types including leukocytes, epithelial cells, endothelial cells, and fibroblasts [25,26]. It is expressed constitutively and abundantly on vascular endothelial cells. Its expression is markedly induced by some pro-inflammatory stimuli like IL-1, TNF-α, interferon-γ (IFN), or LPS [27,28]. Structurally, ICAM-1 is related to the immunoglobulin superfamily of adhesion molecules including VCAM-1, which share a common immunoglobulin-like structure and comprise various

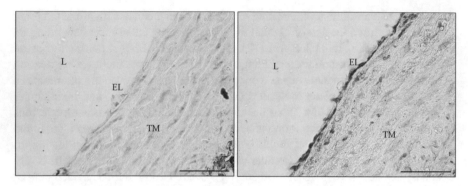

Fig. 8.1 VCAM-1 immunohistochemical staining on aorta. Normotensive rodent exhibited a relatively weaker staining on endothelial lining (EL; *left*) compared to hypertensive rodent, which showed a remarkably intense staining for VCAM-1 (*right*). Abbreviations: *L* lumen, *TM* tunica media. (Magnification: ×400; scale bar: 50 μm)

proteins fundamentally function in both cell-to-cell and cell-to-matrix adherence [29,30]. ICAM-1 contains five extracellular immunoglobulin-like repeat, one transmembrane segment, and a short cytoplasmic tail [30]. A motif within the first and third domains is responsible for binding to the β_2 integrin ligands such as lymphocyte function-associated antigen-1 (LFA-1) expressed on T-lymphocytes and Mac-1 on neutrophil, respectively [31,32]. Through these interactions, ICAM-1 alongside with VCAM-1 plays an important part in the recruitment and trafficking of leukocytes in endothelial cells during inflammatory responses, such as atherosclerosis [33]. While VCAM-1 is crucial during the initial step of leukocytes tethering via VLA-4 ligands, ICAM-1 functions to adhere the leukocytes firmly to the endothelial surface and initiate the process of immune cell transmigratory egress from the circulatory system [33]. Both VCAM-1 and ICAM-1 function to stabilise leukocyte adhesion to the endothelial lining. Endothelial cells, smooth muscle cells, and macrophages are the general sources of VCAM-1 and ICAM-1 expression in atherosclerotic lesions [34,35]. A study demonstrated that VCAM-1, but not ICAM-1, has a considerable role in the formation of early atherosclerotic lesions [36]. However, ICAM-1 was found to mainly involve in the progression of atherosclerosis [37]. Several early studies have reported that the upregulation of VCAM-1 and ICAM-1 is associated with the sites as well as stages of atherosclerotic lesion formation [38–40].

Recruitment of leukocytes to endothelial linings causes damage to the vascular wall. It has been shown recently that ICAM-1 expression was increased significantly in hypertensive rodents ([23]; Fig. 8.2) and in pregnancy-induced hypertensive women [41]. Elevated endothelial expressions of adhesion molecules may indicate a disturbance to the endothelial function [42]. A recent study has reported significant positive correlations of systolic and diastolic blood pressure with ICAM-1 and VCAM-1 and that a decrease in soluble adhesion molecules might suggest the deactivation of endothelium [43]. Therefore, hypertension is considered

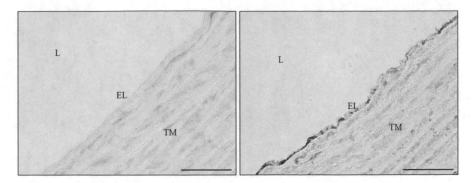

Fig. 8.2 ICAM-1 immunohistochemical staining on aorta. The immunostaining is denser in hypertensive rodent (*right*) than normotensive rodent (*left*). Abbreviations: *EL* endothelial lining, *L* lumen, *TM* tunica media. (Magnification: ×400; scale bar: 50 μm)

as a state of endothelial activation and dysfunction. Both VCAM-1 and ICAM-1 are useful markers for investigating vascular inflammation in hypertension.

Lectin-like Oxidised Low-Density Lipoprotein Receptor-1

Lectin like oxidised low density lipoprotein receptor-1 (LOX-1), a type II membrane protein (52 kD) with a C-terminus lectin-like extracellular domain and a short N-terminus cytoplasmic tail, is a relatively newly discovered receptor. This novel receptor was identified in vascular bovine aortic endothelial cells only a decade ago (Sawamura et al. 1997). It was identified as a major receptor for oxidised low-density lipoprotein (Ox-LDL) on endothelial cells. The C-terminal end residues and several conserved positively charged residues spanning the lectin domain are recognised for Ox-LDL binding [44,45]. It binds, internalises, and degrades the oxidatively modified Ox-LDL, but not native LDL [46]. Therefore, LOX-1 might be an essential mediator for the pathophysiological effects of Ox-LDL in the vascular wall.

Since its discovery, a tremendous number of studies have demonstrated a clearer definition and understanding of the central role played by LOX-1 in atherosclerosis [47–52]. Endothelial dysfunction elicited by Ox-LDL is recognised as the key step in the initiation of atherosclerosis. LOX-1 activation by Ox-LDL causes endothelial functional alterations that are characterised by induction of adhesion molecules [53], monocyte-chemotactic protein [54], smooth muscle cells' proliferation [55], and attenuated endothelium-dependent vasodilatory response by reducing the generation of nitric oxide [56]. Hence, LOX-1 is unquestionably a confirmative marker for atherosclerosis.

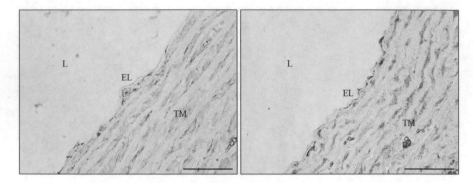

Fig. 8.3 LOX-1 immunohistochemical staining on aorta. Hypertensive rodent demonstrated a relatively stronger staining on endothelial lining (EL; *right*). On the contrary, little or absence of staining on EL in normotensive rodent (*left*). Abbreviations: *L* lumen, *TM* tunica media. (Magnification: ×400; scale bar: 50 μm)

Unexpectedly, however, LOX-1 expression was remarkably upregulated (>20 folds) in the aorta of hypertensive rats, which implies a pathophysiologic role for LOX-1 in hypertension or in hypertensive vascular remodelling [57]. Furthermore, the LOX-1 upregulation in the kidney of Dahl salt-sensitive hypertensive rats was parallel to renal dysfunction, suggesting a possible role of LOX-1 in hypertensive renal disease [58]. We have also found a slight increase in the expression of aortic endothelial LOX-1 in hypertensive rats ([23], Fig. 8.3). LOX-1 may be involved in impairing the endothelium-dependent vasodilatation in these experimental animals. Nonetheless, the LOX-1 expression was found to show a significant direct association with the expressions of VCAM-1 and ICAM-1 in hypertensive group. Inoue et al. [59] have reported that LOX-1 overexpression promoted the expression of VCAM-1 and ICAM-1 on the endothelial cells, which further contributes to vascular inflammation. Hence, it is obvious that LOX-1 could be a novel marker for hypertension. The pathogenetic roles of LOX-1 in hypertension development, however, remain to be determined. Investigating the functions of LOX-1 in hypertension development may represent a novel therapeutic target to control the disease.

P-selectin

The selectin family of adhesion molecules mediates the initial attachment of leukocytes to endothelial surface before their firm adhesion at inflammatory sites. The selectin family consists of three closely associated members sharing common structural features but with differential expression by leukocytes (L-selectin), platelets (P-selectin), and vascular endothelium (P selectin and E-selectin). Among these molecules, P-selectin is of interest because of its unique expression by both platelets and endothelial cells. It is the largest of selectin family with a mass of approximately 140 kDa. P-selectin is also previously known as granule membrane protein 140 (GMP-140) or platelet activation-dependent granule external membrane protein (PADGEM) [60]. It is a cell membrane component of the α granules of platelets (that is how it gets its name) as well as of the Weibel–Palade body (an organelle that stores the coagulation-related von Willebrand factor in endothelial cells). P-selectin is re-localised rapidly from the cytoplasmic bodies to the cell surface upon stimulation by inflammatory mediators [61] as its expression has been observed on the endothelial lining of blood vessel by immunohistochemistry [62].

Structurally, P-selectin consists of an extracellular N-terminus lectin domain, an epidermal growth factor-like region, nine complement regulatory-like protein repeats, a transmembrane section, and a short intracytoplasmic tail [63]. P-selectin binds to its specific mucin-like counter-receptor, designated as P-selectin glycoprotein ligand-1 (PSGL-1) on leukocytes [64], suggesting that it is involved in an interaction between leukocytes and the endothelium, between platelets and leukocytes, and also between platelets and the endothelium themselves [65–67].

Several studies have highlighted the essential role of P-selectin in tethering leukocytes on the endothelial surface and the interaction between platelets and leukocytes [68–70]. Therefore, raised plasma level of P-selectin is considered to be a marker of inflammation. Various experiments have reported raised P-selectin expression in cardiovascular diseases such as stable peripheral artery disease [71], atherosclerosis [72], and unstable angina [73]. Immunohistochemical analysis of human atherosclerotic plaques has demonstrated a strong expression of P-selectin by the endothelial cells overlying active atherosclerotic plaques and fatty streaks [74]. Nevertheless, a study demonstrated a significant and positive correlation of blood pressure and P-selectin that persists in the presence of extreme blood pressure elevation [75]. Increased blood pressure was suggested to be related to an elevated level of plasma P-selectin and endothelial dysfunction (as indicated by high von Willebrand factor levels) [76]. Quite recently, inhibition of P-selection function by inclacumab has been shown to diminish the platelet–leukocyte interaction which may represent an effective treatment for cardiovascular diseases [77]. As it may be involved in and link up various cardiovascular diseases, the expression of P-selectin in hypertension therefore deserves additional studies...

E-selectin

Generally, P-selectin and E-selectin are important in leukocytes recruitment. Both of them are induced at reasonably high density on the vascular endothelial cells at sites of inflammation. Halacheva et al. [78] observed that positive immunostaining for E-selectin on arterial endothelium was accompanied by infiltration of inflammatory cells in the intima. Due to their overlapping function in leukocyte adhesion, P-selectin seems to outshine E-selectin as mice lacking this molecule only exhibit a modest reduction in the atherosclerotic lesion size [79,80]. However, a study has reported the combined effect of these selectins on the development of atherosclerotic lesion [81]. In that study, P- and E-selectin-double-deficient mice were found to develop fatty streaks that were three times smaller than those developed in P-selectin-single-deficient mice.

Unlike ICAM-1 and VCAM-1, E-selectin is expressed only on the endothelial surface after pro-inflammatory stimulation and it can hence be considered a specific marker of endothelial dysfunction in cardiovascular and inflammatory diseases, including hypertension. The imperative involvement of E-selectin in hypertension has been increasingly ascertained by multiple lines of studies [18,82,83]. The possible mechanism played by E-selectin in raising blood pressure may be associated with impaired vasodilatory response and reduced endothelial nitric oxide production [84]. Nonetheless, E-selectin might as well be involved in vascular wall hypertrophy observed in hypertensive animals [85] because it has been previously found to be directly correlated to intima-media thickness in patients with ischemic heart disease [86]. The potential function of E-selectin in hypertension needs to be further elucidated.

Endothelin Receptors

The endothelin system consists of potent endogenous vasoconstrictor peptides that play pivotal roles in vasomotion, vascular smooth muscle cell proliferation, and vascular remodelling. Disturbance in its productions may contribute to elevated peripheral resistance in pathogenesis of hypertension [87]. Endothelin binds to a specific receptor on vascular smooth muscle cells to induce vasoconstriction. Endothelin receptors are classified as subtypes A, B, and C (ET_A, ET_B, and ET_C), respectively [88]. While the knowledge on the subtype C receptor is limited, the ET_A and ET_B receptors have been extensively studied. The ET_A receptors are expressed on vascular smooth muscle cells and function in vasoconstriction and proliferation [89]. Whereby the ET_B receptors predominate on endothelial cells and are present in low densities in vascular smooth muscle cells of aorta, coronary arteries, mesenteric arteries, and veins of laboratory animals and in humans [90]. The ET_B receptors mediate vasodilatation of smooth muscle via the production of nitric oxide and prostacyclin (PGI_2) [91]. Due to its vasodilatory role, the ET_B receptors have been receiving tremendous attentions lately as it may shed light on a potential alternative treatment for diseases involving endothelin disorder [92–94]. Interestingly, the function of ET_B receptors may switch from vasodilatation mode (vasorelaxing phenotype, termed as ET_{B1}) to vasoconstriction mode (contractile phenotype, termed ET_{B2}) under certain pathophysiological conditions [95]. For instance, inflammatory cytokine TNF-α has been recently found to stimulate the vascular smooth muscle cell expression of the contractile ET_{B2} receptors in rodents via activating NF-κB pathway [96]. Immunohistochemical staining on paraffin-embedded arterial sections demonstrated upregulation of ET_{B2} receptors in tunica media after TNF-α

intervention [96]. The novel function of these endothelin receptors represents an interesting study in hypertension.

Vascular Endothelial Growth Factor

Vascular endothelial growth factor (VEGF) is a potent endothelial cell mitogen and fundamental regulator of physiological and pathological angiogenesis. In rodent's circulatory system, VEGF expression is demonstrated by immunohistochemistry on cardiac myocytes, aortic endothelium, and the tunica media of inferior vena cava [97]. VEGF regulates endothelial cell proliferation and migration [98]. Furthermore, it has been shown to induce endothelium-dependent vasodilation, leading to hypotensive effect [99]. VEGF signalling is mediated through interaction with two transmembrane endothelial cell receptor tyrosine kinases: VEGF receptors (VEGFR) types 1 (Flt-1) or VEGFR 2 (KDR). Binding of VEGF to its receptors causes intracellular phosphorylation such as of endothelial nitric oxide synthase (eNOS) resulting in an increased production of nitric oxide, which acts directly on endothelial cells and eventually leads to endothelial cell proliferation, permeability, and blood vessel formation [100]. A study has demonstrated that although the vasodilation and hypotensive effect of VEGF involves both receptors, the KDR is the major receptor mediating this effect [101]. Blood pressure-lowering effect of VEGF has been shown to be accompanied by upregulation of nitric oxide level [102]. Inhibiting the VEGF pathway may lead to hypertension [103]. However, there are studies reporting elevated VEGF expression in both hypertensive human and experimental rats [104–106]. It has been suggested that aside from its central role in angiogenesis, VEGF may also be involved in inflammatory processes of the vascular endothelial cells likely attributable to increased vascular permeability [107]. VEGF may possess multifaceted roles [108]. Investigations into the role played by VEGF in hypertension may lead to exciting findings.

Tumour Necrosis Factor Alpha

Tumour necrosis factor alpha (TNF-α) is another vascular inflammation marker that may be applicable for immunohistochemistry to study the relationship between inflammation and hypertension. It is a pro-inflammatory glycoprotein peptide hormone that plays a significant role in the initial activation of the immunological responses. Recent data have demonstrated the association between an elevated expression of TNF-α and raised blood pressure as well as cardiac dysfunction [109–111], suggesting an important pro-hypertensive role of inflammatory process in the vascular wall. Furthermore, blockage of TNF-α has been shown by immunohistochemical approach to reduce endothelial and smooth muscle cell TNF-α protein expression, which subsequently improves haemodynamic properties and attenuates inflammation in rodents [112]. The exact mechanisms by which TNF-α influences hypertension are still uncertain, but TNF-α has been reported to downregulate endothelial nitric oxide synthase (eNOS) mRNA expression [113], hence causing an inverse relationship between TNF-α and nitric oxide serum levels in rats [111]. Vascular inflammation induced by TNF-α is associated with impaired endothelium-dependent vasodilatory response [114].

Inducible Nitric Oxide Synthase

Nitric oxide is an important endogenous vasodilator in cardiovascular system that regulates systemic and local blood flow. It is synthesised by the catalytic action of a small family of enzymes known as nitric oxide synthase (NOS), comprising of three isoforms: neuronal (nNOS), endothelial (eNOS), and inducible NOS (iNOS) [115]. In normal cardiovascular function, nitric oxide is produced predominantly by eNOS to play a central role in preserving physiological endothelial functions. However, in pathological conditions such as inflammation, iNOS, which is expressed mostly in smooth muscle cells, is stimulated to produce nitric oxide a thousand times more than that produced by eNOS [116,117]. While the relatively low amounts of nitric oxide by eNOS function to protect the cardiovascular system, this exaggerated release of nitric oxide via iNOS has been linked to the production of deleterious oxidative products [118]. Excessive amounts of oxidative products cause detrimental cardiovascular effects, which may be attributable to the reaction between superoxide and nitric oxide-forming peroxynitrite, subsequently promoting the so-called nitrosative stress and endothelial dysfunction [119,120]. A great number of recent reports have pointed out the significant role of iNOS expression in decreased nitric oxide bioavailability, increased oxidative stress, cell damage, impaired endothelial function, inflammatory activation, and eventually the development of hypertension [115,118,121–123]. Moreover, a polymorphism in *iNOS* gene which is associated with predisposition to hypertension and responsiveness to antihypertensive therapy has been identified in human subjects recently [124], further showing the important role of derangement in nitric oxide in the pathophysiology of hypertension. Hence, iNOS is a very useful marker for hypertension, inflammation, and oxidative stress.

Angiotensin II and Its Receptor

Angiotensin II (Ang II) in hypertension is imperative because of its pluripotent properties. Ang II is the major bioactive component of the renin-angiotensin system produced locally in several tissues, including vascular wall, with a central role in the regulation of blood pressure, vasomotion, and sodium and water balance [125,126]. Furthermore, a recent study has reported that Ang II-induced hypertension involves the upregulation of lipogenic gene expression and dysfunction of both the systolic and diastolic cardiac performance in rats [127]. Aside from its involvement in blood pressure regulation, several studies have revealed the pro-inflammatory-inducing effects of Ang II in vascular endothelial cells [128,129]. Inhibition of angiotensin-converting enzyme (ACE) has been shown to reduce the Ang II-induced expressions of inflammatory cytokines [130,131]. In addition, Ang II also increases the generation of intracellular reactive oxygen species in endothelial cells, leading to excessive oxidative stress that may impair endothelial function [132]. This pro-oxidative effect of Ang II is conferred through the Ang II type I (AT_1) receptor-nicotinamide adenine dinucleotide phosphate-oxidase (NADPH) pathway induction of reactive oxygen species production [133–135]. AT_1 receptors are absolutely required by Ang II in the development of hypertension and cardiac hypertrophy [136]. Due to their adjacent expressions, immunohistochemistry may be a useful method to study the local effects of Ang II and AT_1 receptor.

Principles of Immunohistochemistry

Considering the increasingly strong footing of the role of vascular wall inflammation in hypertension, immunohistochemistry has become an important method to investigate the inflammatory responses in circulatory system. Immunohistochemistry is a method used to recognise the presence and location of constituents of tissues (as antigens) in situ by means of corresponding labelled antibodies. The principle of immunohistochemistry lies on the immunoreaction between antibodies to bind specifically to their corresponding antigens. Since antibodies are specific, they will only bind to the antigens of interest in the tissue. The resulting reaction is visualised by the antibodies which are pre-labelled either with a substance that absorbs or emits light (fluorescent detection) or a substance that produces distinct colour (chromogenic detection). Therefore, immunohistochemistry is actually a general term referring to a group of immunostaining techniques including immunofluorescence techniques and immunoperoxidase techniques, in which specially labelled antibodies are applied for detecting proteins of interest.

Although immunohistochemistry is quantitatively less sensitive than other immunoassays such as protein Western blotting or enzyme-linked immunosorbent assay (ELISA), it however gives an overview of an intact tissue in the aspect of the localisation of a particular protein (i.e. in nucleus, cellular membrane, or cytoplasm). Since it is done on a certain tissue rather than the measuring of inflammatory biomarker levels in serum by other methods, say ELISA, immunohistochemistry is like "talking to the cells" and thus the expressions of inflammatory biomarkers may indicate what they are doing in vivo. The results are most likely reflecting what is going on in the vascular wall itself. It provides information about the presence and activation of a protein in a tissue, and it therefore helps to make sense of the numerical data obtained from other methods. Furthermore, due to its wide availability, lower cost, and easy and long preservation of the stained slides, immunohistochemistry has become an important method in the assessment of pathologies [137]. Only a small amount of tissue is needed to detect various parameters in immunohistochemistry. In addition, the continuing advance in computer-aided image analysis has shifted immunohistochemistry from a traditional "merely descriptive" technique to a quantitative measurement. Digital image evaluation has enabled measurements of protein activations by quantifying the colour density of positive immunostaining on a tissue. Therefore, immunohistochemistry is advantageous in providing not only qualitative but also semi-quantitative or quantitative information of protein expressions. Immunohistochemistry selectively detects specific proteins in tissue and has become a routine technique for diagnostic and assessment purposes of certain diseases. Moreover, it is widely applied for experimental purposes to unravel the pathophysiologic mechanisms.

Technical Aspects

Scrupulous attention to the technical details and components when conducting immunohistochemical staining promises easily interpretable slides. From tissue fixation, to antigen retrieval, to staining procedure, every aspect is crucial in determining the outcome.

Specimen Handling

In the circulatory system of a rodent, aside from the heart, aorta is probably a more suitable tissue to be used in studying vascular pathologies due to its larger size and easier excision. From aortic arch, to thoracic aorta, to abdominal aorta, it can be used in immunohistochemical procedure [23,138–140]. Of course, other tissues from the circulatory system such as penile artery [141] or cerebral blood vessel [142] are applicable as well, depending on the aim of an experiment. The very first step in tissue handling is paramount in governing the quality of immunohistochemical results. Necrotic degradation starts immediately once the tissue is separated from its source of nutrients. Hence vascular specimen must be processed properly and rapidly isolated from adherent adipose and connective tissues after harvesting from the animal. Specimen should then be trimmed and cut for fixation. Size of specimen is important. For instance, aorta should be cut into short length so that the fixative can penetrate efficaciously into the tissue. The shorter or thinner the tissue, the faster the fixation can occur. Besides, fixative should be made sure that it is always fresh. After fixation, the specimen is paraffin embedded either longitudinally or horizontally. To study the expressions of vascular wall proteins, longitudinal embedding is recommended. Horizontal embedding is usually done to investigate the entire length of elastic lamellae or vascular smooth muscle cells. Extra care must be taken when longitudinally embedding a blood vessel sample to ensure an approximate 90° to the cutting surface. Serious degree diversion may lead to uneven thickness of the section. Next, the paraffin block is cut in a microtome to a desired thickness. The ideal section should be accomplished at a thickness of 3–5 μm, with a sharp knife blade coupled with smooth and swift cutting technique. Over thick sections, as well as sections with tears, folds, and knife nicks, may lead to false-positive or false-negative interpretations. Tissue sections should be adhered to poly-lysine pre-coated slides to avoid detachment during heat-induced epitope retrieval.

Fixation

Appropriate fixation is a vital determinant for the success of immunohistochemistry. Fixation terminates any ongoing biochemical reactions in the tissues. In addition, fixation also preserves the localisation as well as structure of the antigen to provide a target for antibodies that will be applied to detect the antigen. Neutral-buffered formalin is the most commonly and practically used fixative for routine tissue sections. Other fixatives include paraformaldehyde, Bouin's solution, or HEPES glutamic acid buffer-mediated organic solvent protection effect (HOPE) fixative. Formalin fixes tissues by covalently cross-linking protein amino acid residues by methylene bridges. Briefly, formalin reacts with a peptide sequence to form a methylol (hydroxymethyl) group on an amino acidic residue like arginine, cysteine, lysine, tryptophan, tyrosine, serine, asparagine, or glutamine. Imine groups (also known as Schiff's bases) are then formed by condensation of methylol groups on lysine residues. Due to its instability, imine groups condense with other components such as phenolic, cysteine, or tyrosine to form methylene bridges [143]. These cross-linkages however induce conformational alteration to the protein secondary, tertiary, and quaternary structures and block the accessibility of targeted antigenic sites to the primary antibodies. Therefore, some form of antigenicity restoration is required to increase the sensitivity of immunohistochemical markers in fixed tissues. We further discuss antigen retrieval techniques in the next section.

Optimal fixation time depends on the size of the specimen. Generally, fixation for between 18 and 24 h is recommendable for most application. Under-fixation or delayed fixation causes a gradation of immunohistochemical staining such that there is strong colourisation on the periphery yet weak signal in the core of the tissue section. This might be explained by the adequate fixation near the outside of the specimen, but that the inside of the specimen is preserved only by coagulative fixation from ethanol which would occur during tissue processing [144]. Under-fixation leads to a decrease in reliability of the immunohistochemical staining [145]. On the other side, over-fixation exaggerates the intra- and intermolecular linking that can cause poor epitope presentation, subsequently reducing immunoreactivity between antibodies and antigens [146]. Although over-fixation can mask antigenic sites, antigen retrieval can help to overcome this loss of antigen staining. In contrast, it is irretrievable to rectify under-fixed tissue and may lead to unreliable immunohistochemical results. Practically speaking, it is therefore recommended that even small specimens should be fixed for at least 6–12 h [147].

Antigen Retrieval

Most vascular specimens are fixed in formalin and subsequently embedded in paraffin. While fixation preserves tissue morphology excellently, it also modifies the structure of tissue proteins, therefore masking the targeted antigenic sites. The masking effect by fixation poses a serious obstacle for immunohistochemical detection of the antigen. There were attempts to find the alternative fixatives to replace formalin decades ago, but it seems that an ideal fixative will never be found [148]. Fortunately, the loss of antigenicity can be corrected by various methods in antigen retrieval. The general principle of antigen retrieval lies upon the breaking of protein cross-links which are induced by fixation process, with the subsequent exposition of antigenic sites in order to allow the primary antibodies to bind. Antigen retrieval is advantageous in enhancing the sensitivity of the immunohistochemical detection of epitopes. The increased immunohistochemical sensitivity allows the use of higher antibody dilutions, which lowers not only the cost but the undesirable non-specific background staining on the tissue section. It therefore minimises the possibility of false-negative results and increases the reproducibility as well as the accuracy of the interpretations. Restoration of antigens can be achieved either by physical (such as enzymatically proteolytic digestion, denaturation, or oxidising treatment) or chemical approaches (including the widely employed heat-induced epitope retrieval, or HIER). Although both approaches work by cleaving the cross-links, the choice of method depends on the antigen of interest and/or antibody to be used. It is the responsibility of the individual laboratory to compare and determine the optimal retrieval protocol for individual antigen in tissues before the staining process is carried out in bulk.

Proteolytic treatment of paraffin-embedded section was first introduced by Huang et al. [149] to detect hepatitis B core and surface antigens. Enzymatic antigen retrieval employs proteolytic enzymes to break the methylene bridges. Commonly used proteolytic enzymes include trypsin, pepsin, and pronase, but papain, bromelin, proteinase K, and collagenase are less frequently used. Due to the use of proteolytic enzymes, this retrieval method can sometimes damage the morphology as well as the targeted antigen of the tissue section. So, the kind and concentration of enzyme, and the treatment time and temperature, need to be optimised. Although enzymatic retrieval does not usually recover immunoreactions dramatically compared to heat treatment, but it is still a useful way for antigens that lose their antigenicity when exposed to heat [150].

Nowadays, heat-induced antigen retrieval is the most commonly used method. Introduced in response to the need for a more effective method to unmask the epitopes in the 1990s, this "revolutionarily new" technique yielded satisfactory outcomes and it was widely employed in pathology [151]. Heat-induced antigen retrieval can be performed using a pressure cooker, a scientific microwave, a vegetable steamer, an autoclave, or in water bath. Volkin and Klibanov [152] suggested that heat would reverse the protein conformation altered by fixation, inducing a "renaturation" of the epitopes. Since the mechanism of heat-induced antigen retrieval relies on the cleavage of the cross-links and in the subsequent structural reconstruction of the antigen, all the factors involved in the procedure, which can potentially influence the antigenicity restoration, need to be controlled to guarantee the retrieval efficacy. The factors include pH, molarity, and type of the retrieval buffer, dilution of primary antibody, and treatment time. The most important factor in retrieval efficiency is the heat. Most protocols suggest antigen retrieval time for 10–20 min at about boiling temperature. However, it is only a suggested time. Under-retrieval may lead to weak staining, but over-retrieval may cause excessive non-specific background staining as well as increase the occurrence of tissue detachment from the slides, especially when it is a minute aortic section.

The next influential factor in antigen retrieval would be the choice of retrieval buffer. Generally, sodium citrate buffer at acidic pH 6.0 and Tris-EDTA buffer at basic pH 9.0 are the most popularly used retrieval solutions for the majority of antigens. Although there is no universally optimal pH for antigen retrieval and the pH choice depends on the targeted antigen, a study has elegantly demonstrated the important role of retrieval solution pH in the process of retrieving antigen [153]. The generation of electrostatic repulsion and hydrophobic attractive forces between antigen molecules in acidic or basic medium might explain the improved immunoreaction. Antigen extends and exposes its hydrophobic and hydrophilic portions after the destruction of methylene bridges by heat (Emoto et al. 2005). This is the very moment where pH fundamentally influences the outcome of the retrieval process. At neutral pH, the antigens are entangled together due to the simultaneous presence of hydrophobic and ionic forces. At acidic or basic pH however, the antigens are charged negatively or positively to create two opposing forces that cooperatively prevent entangling of the protein and thus "open up" the antigenic sites. Kim et al. [153] found that increasing buffer pH from 6.0 to 9.0 yielded gradually improved retrieved antigen immunoreactivity against most antibodies. This may be due to the fact that acidic hydrolysis is much more complex and has a more rate-limiting step than basic hydrolysis. Therefore, in order to avoid under- or over-retrieval, we recommend a control experiment beforehand to evaluate the optimal retrieval protocol for a particular primary antibody, where a known positive tissue control for that antibody is retrieved in either sodium citrate or Tris-EDTA buffer for a range of time at 5 min intervals. At the same time, different dilutions of the antibody are tested. An optimised antibody dilution should produce the highest colour intensity of positive staining with the non-specific background staining kept at the lowest. It is always advisable to start by using the antibody at the dilution recommended on the supplier's technical datasheet.

Briefly, the optimisation method for mouse monoclonal anti-rat intercellular adhesion molecule-1 (ICAM-1) antibody as an example is discussed here. Human tonsil is the positive tissue control for ICAM-1 as recommended on the product datasheet. Even after being microwave heated for more than 10 min in sodium citrate buffer at pH 6.0, the brownish staining in lymphatic nodules, where ICAM-1 is abundantly expressed, was still absent or weak (Fig. 8.4a–c). This might be due to the low concentration of antibody (1:100) that had been applied. So, the dilution

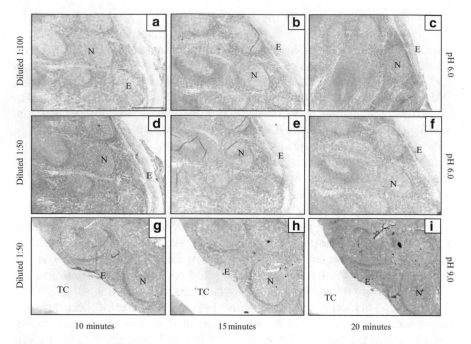

Fig. 8.4 Photomicrographs showing immunohistochemical optimisation for mouse monoclonal antibody anti-rat ICAM-1 on human tonsil sections. The sections are incubated with antibody at dilution factor of 1:100 or 1:50 for an hour at room temperature after being heat-retrieved in either sodium citrate pH 6.0 buffer or Tris-EDTA pH 9.0 buffer, for 10, 15 or 20 minutes. Brownish colour in lymphatic nodules (N) indicates positive staining for ICAM-1 antigens. The optimal condition is chosen based on the colourisation. The sample retrieved in pH 6.0 buffer (**a** & **b**) does not react well with the antibody (diluted 1:100) until it is heated for 20 minutes (**c**). When the concentration of the antibody is increased, it produces satisfactory colour after only a 15 minutes retrieval (**d**). However, the reaction becomes poorer when the sample is heated longer than that (**e** & **f**). When the pH of the retrieval buffer increases to pH 9.0, it gives good staining after 10 or 15 minutes heating (**g** & **h**). But heating for 20 minutes yields oversaturation colour (**i**). Abbreviations: E, epithelium; TC, tonsillar crypt. (Magnification: 4×; scale bar: 500 μm)

of antibody was decreased to 1:50. However, the staining was still weak regardless of how long the sections were heated (Fig. 8.4d–f). At this point, the results suggested that pH 6.0 might not be suitable for the antibody and/or antigen. Then, the sections were retrieved in Tris-EDTA pH 9.0 buffer for various durations. Results showed that retrieval for 20 min caused the tonsil section to be over-retrieved, leading to colour over-saturation and undesirable background staining around the lymphatic nodules (Fig. 8.4i). Surprisingly, retrieval for 15 min constantly gave a rather weaker staining compared to 10 min (Fig. 8.4h). Retrieval for 10 min produced satisfactory staining (Fig. 8.4g). The results were in agreement with a previous study by Kim et al. [153] reporting that basic buffers yielded more satisfactory immunostaining than the acidic buffers. These findings also emphasise the importance of optimising the pH of retrieval solution before immunostaining the samples. Table 8.1 shows the suggested optimal conditions for different antibodies. Please note that

Table 8.1 Suggestions of optimal conditions for immunohistochemical staining of inflammatory biomarkers

Antibody (anti-)	Fixative	Retrieval buffer	Antibody dilution	Heat-induced retrieval time	Tissue	References
VCAM-1	10 % neutral-buffered formalin	Sodium citrate pH 6.0	1:100	10 min	Aorta	Ng et al. [23]
ICAM-1	10 % neutral-buffered formalin	Tris-EDTA pH 9.0	1:50	10 min	Aorta	Ng et al. [23]
LOX-1	10 % neutral-buffered formalin	Sodium citrate pH 6.0	1:100	10 min	Aorta	Ng et al. [23]
P-selectin	4 % paraformaldehyde	0.037 % EDTA/ 0.055 % Tris buffer	1:50	5 min	Artery	Dever et al. [68]
E-selectin	10 % formalin	EDTA buffer pH 8.0	1:100	10 min	Testes	Sukhotnik et al. [154]
ET$_A$ receptor	10 % buffered formalin	Citrate buffer pH 6.0	1:100	10 min	Pulmonary artery	Ishida et al. [155]
ET$_B$ receptor	Paraformaldehyde	Citric acid buffer pH 6.0	1:200	20 min	Mesenteric artery	Zhang et al. [96]
TNF-α	Buffered formalin	Citrate buffer pH 6.0	1:200	18 min	Plantar section	Nobre et al. [156]
VEGF	10 % phosphate-buffered paraformaldehyde	Citrate buffer pH 6.0	–	15 min	Heart	Zaitone and Abo-Gresha [157]
iNOS	10 % phosphate-buffered paraformaldehyde	Citrate buffer pH 6.0	–	15 min	Heart	Zaitone and Abo-Gresha [157]
Ang II	Formalin	Citrate buffer pH 6.0	1:500	5 min (for three times)	Ovary	Harata et al. [158]
AT$_1$R	4 % paraformaldehyde	Citrate buffer pH 6.0	1:150	–	Brain	Thomas and Lemmer [159]

they are only suggested methods. Individual laboratory needs to find out the optimal retrieval protocol for individual antigen in certain tissues before the staining process is carried out in large amount.

Immunohistochemical Staining

There are numerous immunohistochemical staining techniques that vary from simple one-step procedures, where the label is directly linked to the primary antibody, i.e. direct methods, to complex multiple-step procedures, where the label is conjugated to a secondary antibody or a new micropolymer labelling technology, i.e. indirect methods. Although indirect methods are more complicated and consume more time than direct methods, the indirect methods are recommended on vascular tissues because the methods generally yield very sensitive and reproducible results. Other researchers also prefer indirect methods over direct-labelled antibody methods [160,161]. This is especially useful when the protein of interest is expressed low in the specimen. Of the labels conjugated to a secondary detector, horseradish peroxidase (HRP) and alkaline phosphatase are the most commonly used enzymes. A number of chromogens or substrates for these enzymes including 3,3'-diaminobenzidine (DAB), 3-amino-9-ethyl carbazol (AEC), and 5-bromo-4-chloro-3-indoyl phosphate/nitroblue tetrazolium (BCIP/NBT) are used to visualise the antigen–antibody binding. Chromogens DAB and AEC produce intense and contrasting colourisation on the tissues. However, the pitfall in these staining systems would be the endogenous peroxidase activity that can lead to false-positive staining. To avoid this, one should first use hydrogen peroxide to suppress the endogenous peroxidase activity in the tissue. A blocking step also helps to reduce the background staining [161].

During the staining procedures, drying of the tissue sections should not be allowed to occur because drying at any stage will cause non-specific binding, which ultimately leads to frustrating background staining. All incubations, from the blocking step to the colouring step, must be carried out in a humidified chamber. Commercially available chambers or trays with lid provide a constant humid environment for the staining procedures. However, a humidified chamber can actually be built to save cost. A cell culture plate or even a shallow plastic box with a sealed lid and some wet tissue paper pieces will be adequate to make a chamber. To ensure the slides can lay flat to avoid drainage of the incubated reagents, glue a pair of fit-to-the-chamber-size plastic serological pipette tubes at the bottom with a reasonable distance between them. Besides, placing these two parallel pipette tubes with each tip at different sides in the chamber is recommended. This provides a level and an even surface for the slides to rest on and to keep away from the wet tissue papers beneath it.

Parallel control experiment during staining procedures is pivotal to identify false-negative immunostaining. A positive tissue control is recommended to ensure that the antibody is performing as expected. It also aims to check if the staining protocol is properly executed. However, a reliable positive tissue should be appropriately representative with its external or internal tissue components included for analysis and interpretation. So, in the case of investigating vascular inflammation, a normal vascular tissue might not be appropriate to use as a positive control. The choices of positive tissue control can usually be found in the technical datasheet of the primary antibody. For example, the suitable positive tissue control for the inflammatory marker vascular cell adhesion molecule-1 (VCAM-1) is tonsil sample. In addition, a negative control experiment is recommended to identify false-positive result, where the positive tissue is not incubated with antibody.

Nuclear counterstaining may be optional after immunohistochemical staining the tissue. However, counterstaining is strongly recommended in order to visualise nuclei and provide a better contrasting colourisation on the tissue. Some commonly used counterstains include haematoxylin, nuclear fast red, methyl green, toluidine blue, or fluorescent stains. We recommend haematoxylin due to its wide availability and simple preparation. However, it is very important to ensure an appropriate contrast between the immunochemical stain and the nuclear counterstain to avoid visual confusion. Finally, the stained tissues are cover-slipped and mounted with DPX mounting medium. The slides are left to dry in the fume hood before interpretation is carried out.

Quantitative Analysis

Often, immunohistochemically stained vascular specimens are evaluated descriptively by interpreting the immunostaining as either positive or negative based on mere looking for a brownish or reddish colourisation on the tissue. In order to obtain a more quantitative measure (accurately speaking, it is a semi-quantitative technique), the extent and intensity of positive staining on the sections can be evaluated by assigning an arbitrary scale based on appropriate criteria, for example, from the thick and complete stained ring to the patchy and irregular staining of endothelial layer of the aorta for VCAM-1 and ICAM-1 [162]. These approaches usually depend on the observations of multiple investigators who are blind to the grouping. Therefore, the interpretation of such results is very subjective and biased, bringing a substantial methodological inconsistency. It is also limited to the data range that can be generated (i.e. score is usually quantified on a 0–5 scale). Quantitative immunohistochemistry is required to reliably and accurately assess the expression of inflammatory proteins in tissue.

In order to make immunohistochemical studies more quantitative and objective, the continuous developments in computer-assisted biological imaging in the last few years have definitely been revolutionised and have established image analysis techniques to achieve true quantification and extract functional per-cell information from the tissue. The market nowadays provides a vast variety of image analysis software for immunohistochemistry. A great number of studies utilising these technological advance to establish the computerised image analysis protocols for quantifying the colour density or amount of immunostaining signal [163–168]. Some excellent articles on computer-aided immunohistochemistry analysis are recommendable ([169,170, 164]; Mandarim-de-Lacerda et al. 2010). The quantification of the expression of specific proteins via the image analysis is particularly important in providing critical information about certain multifactorial pathologies such as hypertension.

We have established a protocol for quantifying the expression of inflammatory adhesion molecules on endothelial lining of the aorta in hypertensive rodents using image analysis technology based on previous descriptions by Mandarim-de-Lacerda et al. (2010) and Moraes-Teixeira et al. (2010). The protocol is depicted in Fig. 8.5. We used the software Image-Pro Plus version 7.0 (Media Cybernetics, Silver Spring, MD, USA) with Windows 7 to assess the images. First and foremost, digital images

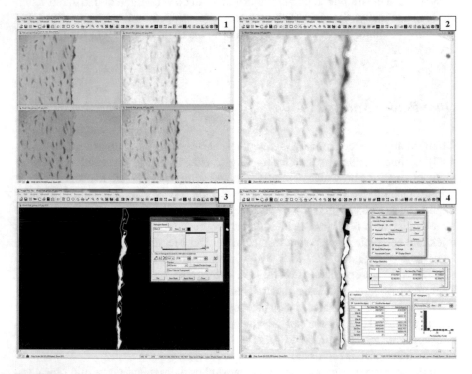

Fig. 8.5 Immunohistochemistry analysis of aortic section. (1) Colour image is thresholded into greyscale images through red, blue, and green channels. The image with distinct immunostaining and light background is selected for further analysis. (2) Then, tunica intima boundary of aorta is delimited by the "area-of-interest" tool. (3) The image is segmented to create a binary image. White colour represents the immunostaining, and black colour represents the unstained background. (4) Lastly, the total area of tunica intima occupied by immunostaining (now it is white) is quantified. Colour density is expressed as the percentage of tunica intima area (%)

of the aorta were acquired (JPEG format, 24-bit colour, 2560 × 1920 pixel resolution, at 200× magnification) with a MicroPublisher 5.0 RTV camera (Q Imaging, Surrey, BC, Canada) connected to a Nikon Eclipse 80i light microscope (Nikon Corporation, Tokyo, Japan). Intima and media of the entire aortic ring were divided into several fields. So, from each specimen, approximately four to seven images could be captured.

The colour images were converted into greyscale pictures by thresholding the images through blue, green, or red colour filter channels. An image with distinct staining area and light background was selected from the three generated images. By doing so, it is more straightforward to select the colour range for the area of interest (i.e. the immunostaining spots). Next, tunica intima boundary was delimited using an irregular "area-of-interest" tool. Aortic specimens are exhausting to delimit mostly due to the fact that two different closely situated tissue components need to be distinguished carefully: the tunica intima and the tunica media. The boundary between tunica intima and tunica media can be distinguished based on delimiting the internal elastic lamella, which is the first and innermost elastic lining in the media. Tunica intima consists of only one-cell lining. Delimitation is recommended to be carried out under higher zooming.

After that, the immunostaining area was segmented into a new binary image, where white colour represented immunostaining and black colour represented unstained area. Then, the percentage of area that was occupied by white colour (i.e. the immunostaining) inside the delimited tunica intima was quantified using the image histogram tool. This tool allows the visualisation of the selected grey values within the region of interest between 0, which represents pure white, and 255 represents pure black. Finally, immunostaining was expressed as the percentage of the entire tunica intima area (%). For accuracy and reproducibility of the result, we recommend that measurements are obtained from five nonconsecutive aortic sections from each rodent.

Avoiding the Pitfalls

From what have been discussed, there are indeed various potential pitfalls in vascular immunohistochemistry that need to be avoided as good as possible. In the pre-staining stage, it is crucial to ensure adequate and immediate fixation of the tissue. Meticulous attention must be taken when longitudinally embedding, in particular a blood vessel sample. In the staining stage, both retrieval and immunostaining techniques must be carefully optimised for each antibody. All incubations must be carried out in a constant humid environment to avoid drying of the section. In the post-staining stage, mounting should be done scrupulously to prevent trapping of air bubbles under the cover slip that can interfere with image analysis. Finally, multiple evaluations of colour density of the immunostaining from each rodent should be executed to guarantee an accurate reading.

Conclusion

The mounting data as discussed suggests the significant role of vascular inflammation in the development of various diseases including hypertension, which may involve a variety of inflammatory markers or bioactive mediators. Immunohistochemistry represents one of the sensitive methods for investigating the inflammatory effects in circulatory system. Further investigations lead to a deeper understanding of the disease pathogenesis as well as potential discovery of novel therapeutic agents.

Acknowledgement The authors would like to acknowledge the financial supports from the Universiti Kebangsaan Malaysia (FF-161-2010 and UKM-GUP-SK-08-21-299 grants). The authors also thank Prof. Dr Siti Aishah Md Ali, Assoc. Prof. Dr Faizah Othman, and Madam Sinar Suriya Muhamad for their technical assistance as well as Ms Xin-Fang Leong for her editorial assistance.

References

1. Harrison DG, Vinh A, Lob H, Madhur MS. Role of the adaptive immune system in hypertension. Curr Opin Pharmacol. 2010;10(2):203–7.
2. Savoia C, Schiffrin EL. Inflammation in hypertension. Curr Opin Nephrol Hypertens. 2006;15(2):152–8.
3. Schiffrin EL. The immune system: role in hypertension. Can J Cardiol. 2012;29:543–8.
4. Khraibi AA, Norman Jr RA, Dzielak DJ. Chronic immunosuppression attenuates hypertension in Okamoto spontaneously hypertensive rats. Am J Physiol. 1984;247(5 Pt 2):H722–6.
5. Norman Jr RA, Dzielak DJ, Bost KL, Khraibi AA, Galloway PG. Immune system dysfunction contributes to the aetiology of spontaneous hypertension. J Hypertens. 1985; 3(3):261–8.
6. Dzielak DJ. The immune system and hypertension. Hypertension. 1992;19(1 Suppl):I36–44.
7. Parissis JT, Korovesis S, Giazitzoglou E, Kalivas P, Katritsis D. Plasma profiles of peripheral monocyte-related inflammatory markers in patients with arterial hypertension. Correlations with plasma endothelin-1. Int J Cardiol. 2002;83(1):13–21.
8. Kampus P, Muda P, Kals J, Ristimäe T, Fischer K, Teesalu R, Zilmer M. The relationship between inflammation and arterial stiffness in patients with essential hypertension. Int J Cardiol. 2006;112(1):46–51.
9. Elmarakby AA, Faulkner J, Posey SP, Sullivan JC. Induction of hemeoxygenase-1 attenuates the hypertension and renal inflammation in spontaneously hypertensive rats. Pharmacol Res. 2010;62(5):400–7.
10. O'Sullivan JB, Ryan KM, Harkin A, Connor TJ. Noradrenaline reuptake inhibitors inhibit expression of chemokines IP-10 and RANTES and cell adhesion molecules VCAM-1 and ICAM-1 in the CNS following a systemic inflammatory challenge. J Neuroimmunol. 2010; 220(1–2):34–42.
11. Van Kampen C, Mallard BA. Regulation of bovine intercellular adhesion molecule 1 (ICAM-1) and vascular cell adhesion molecule 1 (VCAM-1) on cultured aortic endothelial cells. Vet Immunol Immunopathol. 2001;79(1–2):129–38.
12. Chen C, Mobley JL, Dwir O, Shimron F, Grabovsky V, Lobb RR, Shimizu Y, Alon R. High affinity very late antigen-4 subsets expressed on T cells are mandatory for spontaneous adhesion strengthening but not for rolling on VCAM-1 in shear flow. J Immunol. 1999; 162(2):1084–95.

13. Choi SE, Jang HJ, Kang Y, Jung JG, Han SJ, Kim HJ, Kim DJ, Lee KW. Atherosclerosis induced by a high-fat diet is alleviated by lithium chloride via reduction of VCAM expression in ApoE-deficient mice. Vascul Pharmacol. 2010;53(5–6):264–72.
14. Tikellis C, Jandeleit-Dahm KA, Sheehy K, Murphy A, Chin-Dusting J, Kling D, Sebokova E, Cooper ME, Mizrahi J, Woollard KJ. Reduced plaque formation induced by rosiglitazone in an STZ-diabetes mouse model of atherosclerosis is associated with downregulation of adhesion molecules. Atherosclerosis. 2008;199(1):55–64.
15. Wu Y, Zhang R, Zhou C, Xu Y, Guan X, Hu J, Xu Y, Li S. Enhanced expression of vascular cell adhesion molecule-1 by corticotrophin-releasing hormone contributes to progression of atherosclerosis in LDL receptor-deficient mice. Atherosclerosis. 2009;203(2):360–70.
16. Tsoyi K, Kim WS, Kim YM, Kim HJ, Seo HG, Lee JH, Yun-Choi HS, Chang KC. Upregulation of PTEN by CKD712, a synthetic tetrahydroisoquinoline alkaloid, selectively inhibits lipopolysaccharide-induced VCAM-1 but not ICAM-1 expression in human endothelial cells. Atherosclerosis. 2009;207(2):412–9.
17. Boulbou MS, Koukoulis GN, Makri ED, Petinaki EA, Gourgoulianis KI, Germenis AE. Circulating adhesion molecules levels in type 2 diabetes mellitus and hypertension. Int J Cardiol. 2005;98(1):39–44.
18. Glowinska B, Urban M, Peczynska J, Florys B. Soluble adhesion molecules (sICAM-1, sVCAM-1) and selectins (sE selectin, sP selectin, sL selectin) levels in children and adolescents with obesity, hypertension, and diabetes. Metabolism. 2005;54(8):1020–6.
19. Lozano-Nuevo JJ, Estrada-Garcia T, Vargas-Robles H, Escalante-Acosta BA, Rubio-Guerra AF. Correlation between circulating adhesion molecules and resistin levels in hypertensive type-2 diabetic patients. Inflamm Allergy Drug Targets. 2011;10(1):27–31.
20. Cachofeiro V, Miana M, de las Heras N, Martín-Fernandez B, Ballesteros S, Balfagon G, Lahera V. Inflammation: a link between hypertension and atherosclerosis. Curr Hypertens Rev. 2009;5(1):40–8.
21. Costanzo A, Moretti F, Burgio VL, Bravi C, Guido F, Levrero M, Puri PL. Endothelial activation by angiotensin II through NFkappaB and p38 pathways: Involvement of NFkappaB-inducible kinase (NIK), free oxygen radicals, and selective inhibition by aspirin. J Cell Physiol. 2003;195(3):402–10.
22. Tummala PE, Chen XL, Sundell CL, Laursen JB, Hammes CP, Alexander RW, Harrison DG, Medford RM. Angiotensin II induces vascular cell adhesion molecule-1 expression in rat vasculature: a potential link between the renin-angiotensin system and atherosclerosis. Circulation. 1999;100(11):1223–9.
23. Ng CY, Kamisah Y, Faizah O, Jaarin K. The role of repeatedly heated soybean oil in the development of hypertension in rats: association with vascular inflammation. Int J Exp Pathol. 2012;93(5):377–87.
24. Norsidah KZ, Asmadi AY, Azizi A, Faizah O, Kamisah Y. Palm Tocotrienol-rich fraction improves vascular proatherosclerotic changes in hyperhomocysteinemic rats. Evid Based Complement Alternat Med. 2013;2013:976967.
25. Dippold W, Wittig B, Schwaeble W, Mayet W, Meyer zum Büschenfelde KH. Expression of intercellular adhesion molecule 1 (ICAM-1, CD54) in colonic epithelial cells. Gut. 1993;34(11):1593–7.
26. Kelly CP, O'Keane JC, Orellana J, Schroy 3rd PC, Yang S, LaMont JT, Brady HR. Human colon cancer cells express ICAM-1 in vivo and support LFA-1-dependent lymphocyte adhesion in vitro. Am J Physiol. 1992;263(6 Pt 1):G864–70.
27. Dustin ML, Rothlein R, Bhan AK, Dinarello CA, Springer TA. Induction by IL 1 and interferon-gamma: tissue distribution, biochemistry, and function of a natural adherence molecule (ICAM-1). J Immunol. 1986;137(1):245–54.
28. Pober JS, Bevilacqua MP, Mendrick DL, Lapierre LA, Fiers W, Gimbrone Jr MA. Two distinct monokines, interleukin 1 and tumor necrosis factor, each independently induce biosynthesis and transient expression of the same antigen on the surface of cultured human vascular endothelial cells. J Immunol. 1986;136(5):1680–7.

29. Hopkins AM, Baird AW, Nusrat A. ICAM-1: targeted docking for exogenous as well as endogenous ligands. Adv Drug Deliv Rev. 2004;56(6):763–78.
30. Staunton DE, Marlin SD, Stratowa C, Dustin ML, Springer TA. Primary structure of ICAM-1 demonstrates interaction between members of the immunoglobulin and integrin supergene families. Cell. 1988;52(6):925–33.
31. Staunton DE, Dustin ML, Erickson HP, Springer TA. The arrangement of the immunoglobulin-like domains of ICAM-1 and the binding sites for LFA-1 and rhinovirus. Cell. 1990;61(2):243–54.
32. Vonderheide RH, Tedder TF, Springer TA, Staunton DE. Residues within a conserved amino acid motif of domains 1 and 4 of VCAM-1 are required for binding to VLA-4. J Cell Biol. 1994;125(1):215–22.
33. Springer TA. Adhesion receptors of the immune system. Nature. 1990;346(6283):425–34.
34. Giuriato L, Chiavegato A, Pauletto P, Sartore S. Correlation between the presence of an immature smooth muscle cell population in tunica media and the development of atherosclerotic lesion. A study on different-sized rabbit arteries from cholesterol-fed and Watanabe heritable hyperlipemic rabbits. Atherosclerosis. 1995;116(1):77–92.
35. Jang Y, Lincoff AM, Plow EF, Topol EJ. Cell adhesion molecules in coronary artery disease. J Am Coll Cardiol. 1994;24(7):1591–601.
36. Nachtigal P, Semecky V, Kopecky M, Gojova A, Solichova D, Zdansky P, Zadak Z. Application of stereological methods for the quantification of VCAM-1 and ICAM-1 expression in early stages of rabbit atherogenesis. Pathol Res Pract. 2004;200(3):219–29.
37. Kitagawa K, Matsumoto M, Sasaki T, Hashimoto H, Kuwabara K, Ohtsuki T, Hori M. Involvement of ICAM-1 in the progression of atherosclerosis in APOE-knockout mice. Atherosclerosis. 2002;160(2):305–10.
38. Iiyama K, Hajra L, Iiyama M, Li H, DiChiara M, Medoff BD, Cybulsky MI. Patterns of vascular cell adhesion molecule-1 and intercellular adhesion molecule-1 expression in rabbit and mouse atherosclerotic lesions and at sites predisposed to lesion formation. Circ Res. 1999;85(2):199–207.
39. Nakashima Y, Raines EW, Plump AS, Breslow JL, Ross R. Upregulation of VCAM-1 and ICAM-1 at atherosclerosis-prone sites on the endothelium in the ApoE-deficient mouse. Arterioscler Thromb Vasc Biol. 1998;18(5):842–51.
40. Zibara K, Chignier E, Covacho C, Poston R, Canard G, Hardy P, McGregor J. Modulation of expression of endothelial intercellular adhesion molecule-1, platelet-endothelial cell adhesion molecule-1, and vascular cell adhesion molecule-1 in aortic arch lesions of apolipoprotein E-deficient compared with wild-type mice. Arterioscler Thromb Vasc Biol. 2000;20(10):2288–96.
41. Wilczyński JR, Banasik M, Tchórzewski H, Głowacka E, Malinowski A, Szpakowski M, Lewkowicz P, Wieczorek A, Zeman K, Wilczyński J. Expression of intercellular adhesion molecule-1 on the surface of peripheral blood and decidual lymphocytes of women with pregnancy-induced hypertension. Eur J Obstet Gynecol Reprod Biol. 2002;102(1):15–20.
42. Iliodromitis EK, Andreadou I, Markantonis-Kyroudis S, Mademli K, Kyrzopoulos S, Georgiadou P, Kremastinos DT. The effects of tirofiban on peripheral markers of oxidative stress and endothelial dysfunction in patients with acute coronary syndromes. Thromb Res. 2007;119(2):167–74.
43. Tadzic R, Mihalj M, Vcev A, Ennen J, Tadzic A, Drenjancevic I. The effects of arterial blood pressure reduction on endocan and soluble endothelial cell adhesion molecules (CAMs) and CAMs ligands expression in hypertensive patients on Ca-channel blocker therapy. Kidney Blood Press Res. 2013;37(2–3):103–15.
44. Chen M, Sawamura T. Essential role of cytoplasmic sequences for cell-surface sorting of the lectin-like oxidized LDL receptor-1 (LOX-1). J Mol Cell Cardiol. 2005;39(3):553–61.
45. Xie Q, Matsunaga S, Shi X, Ogawa S, Niimi S, Wen Z, Tokuyasu K, Machida S. Refolding and characterization of the functional ligand-binding domain of human lectin-like oxidized LDL receptor. Protein Expr Purif. 2003;32(1):68–74.

46. Hu B, Li D, Sawamura T, Mehta JL. Oxidized LDL through LOX-1 modulates LDL-receptor expression in human coronary artery endothelial cells. Biochem Biophys Res Commun. 2003;307(4):1008–12.
47. Chen M, Masaki T, Sawamura T. LOX-1, the receptor for oxidized low-density lipoprotein identified from endothelial cells: implications in endothelial dysfunction and atherosclerosis. Pharmacol Ther. 2002;95(1):89–100.
48. Ishino S, Mukai T, Kume N, Asano D, Ogawa M, Kuge Y, Minami M, Kita T, Shiomi M, Saji H. Lectin-like oxidized LDL receptor-1 (LOX-1) expression is associated with atherosclerotic plaque instability—analysis in hypercholesterolemic rabbits. Atherosclerosis. 2007; 195(1):48–56.
49. Ishiyama J, Taguchi R, Yamamoto A, Murakami K. Palmitic acid enhances lectin-like oxidized LDL receptor (LOX-1) expression and promotes uptake of oxidized LDL in macrophage cells. Atherosclerosis. 2010;209(1):118–24.
50. Kume N, Kita T. Lectin-like oxidized low-density lipoprotein receptor-1 (LOX-1) in atherogenesis. Trends Cardiovasc Med. 2001;11(1):22–5.
51. Uchida K, Suehiro A, Nakanishi M, Sawamura T, Wakabayashi I. Associations of atherosclerotic risk factors with oxidized low-density lipoprotein evaluated by LOX-1 ligand activity in healthy men. Clin Chim Acta. 2011;412(17–18):1643–7.
52. White SJ, Sala-Newby GB, Newby AC. Overexpression of scavenger receptor LOX-1 in endothelial cells promotes atherogenesis in the ApoE(−/−) mouse model. Cardiovasc Pathol. 2011;20(6):369–73.
53. Lii CK, Lei YP, Yao HT, Hsieh YS, Tsai CW, Liu KL, Chen HW. Chrysanthemum morifolium Ramat. reduces the oxidized LDL-induced expression of intercellular adhesion molecule-1 and E-selectin in human umbilical vein endothelial cells. J Ethnopharmacol. 2010; 128(1):213–20.
54. Jang MK, Kim JY, Jeoung NH, Kang MA, Choi MS, Oh GT, Nam KT, Lee WH, Park YB. Oxidized low-density lipoproteins may induce expression of monocyte chemotactic protein-3 in atherosclerotic plaques. Biochem Biophys Res Commun. 2004; 323(3):898–905.
55. Chahine MN, Blackwood DP, Dibrov E, Richard MN, Pierce GN. Oxidized LDL affects smooth muscle cell growth through MAPK-mediated actions on nuclear protein import. J Mol Cell Cardiol. 2009;46(3):431–41.
56. Novella S, Laguna-Fernández A, Lázaro-Franco M, Sobrino A, Bueno-Betí C, Tarín JJ, Monsalve E, Sanchís J, Hermenegildo C. Estradiol, acting through estrogen receptor alpha, restores dimethylarginine dimethylaminohydrolase activity and nitric oxide production in oxLDL-treated human arterial endothelial cells. Mol Cell Endocrinol. 2013;365(1):11–6.
57. Nagase M, Hirose S, Sawamura T, Masaki T, Fujita T. Enhanced expression of endothelial oxidized low-density lipoprotein receptor (LOX-1) in hypertensive rats. Biochem Biophys Res Commun. 1997;237(3):496–8.
58. Nagase M, Kaname S, Nagase T, Wang G, Ando K, Sawamura T, Fujita T. Expression of LOX-1, an oxidized low-density lipoprotein receptor, in experimental hypertensive glomerulosclerosis. J Am Soc Nephrol. 2000;11(10):1826–36.
59. Inoue K, Arai Y, Kurihara H, Kita T, Sawamura T. Overexpression of lectin-like oxidized low-density lipoprotein receptor-1 induces intramyocardial vasculopathy in apolipoprotein E-null mice. Circ Res. 2005;97(2):176–84.
60. Kansas GS. Selectins and their ligands: current concepts and controversies. Blood. 1996;88(9):3259–87.
61. Wagner DD. The Weibel-Palade body: the storage granule for von Willebrand factor and P-selectin. Thromb Haemost. 1993;70(1):105–10.
62. Ortiz-Rey JA, Suárez-Peñaranda JM, San Miguel P, Muñoz JI, Rodríguez-Calvo MS, Concheiro L. Immunohistochemical analysis of P-Selectin as a possible marker of vitality in human cutaneous wounds. J Forensic Leg Med. 2008;15(6):368–72.

63. Johnston GI, Cook RG, McEver RP. Cloning of GMP-140, a granule membrane protein of platelets and endothelium: sequence similarity to proteins involved in cell adhesion and inflammation. Cell. 1989;56(6):1033–44.
64. Sako D, Chang XJ, Barone KM, Vachino G, White HM, Shaw G, Veldman GM, Bean KM, Ahern TJ, Furie B, Cumming DA, Larsen GR. Expression cloning of a functional glycoprotein ligand for P-selectin. Cell. 1993;75(6):1179–86.
65. Alon R, Hammer DA, Springer TA. Lifetime of the P-selectin-carbohydrate bond and its response to tensile force in hydrodynamic flow. Nature. 1995;374(6522):539–42.
66. Geng JG, Bevilacqua MP, Moore KL, McIntyre TM, Prescott SM, Kim JM, Bliss GA, Zimmerman GA, McEver RP. Rapid neutrophil adhesion to activated endothelium mediated by GMP-140. Nature. 1990;343(6260):757–60.
67. Hamburger SA, McEver RP. GMP-140 mediates adhesion of stimulated platelets to neutrophils. Blood. 1990;75(3):550–4.
68. Dever GJ, Benson R, Wainwright CL, Kennedy S, Spickett CM. Phospholipid chlorohydrin induces leukocyte adhesion to ApoE$^{-/-}$ mouse arteries via upregulation of P-selectin. Free Radic Biol Med. 2008;44(3):452–63.
69. Shodai T, Suzuki J, Kudo S, Itoh S, Terada M, Fujita S, Shimazu H, Tsuji T. Inhibition of P-selectin-mediated cell adhesion by a sulfated derivative of sialic acid. Biochem Biophys Res Commun. 2003;312(3):787–93.
70. Zhao SP, Xu DY. Oxidized lipoprotein(a) enhanced the expression of P-selectin in cultured human umbilical vein endothelial cells. Thromb Res. 2000;100(6):501–10.
71. Blann A, Morris J, McCollum C. Soluble L-selectin in peripheral arterial disease: relationship with soluble E-selectin and soluble P-selectin. Atherosclerosis. 1996;126(2):227–31.
72. Molenaar TJ, Twisk J, de Haas SA, Peterse N, Vogelaar BJ, van Leeuwen SH, Michon IN, van Berkel TJ, Kuiper J, Biessen EA. P-selectin as a candidate target in atherosclerosis. Biochem Pharmacol. 2003;66(5):859–66.
73. Atalar E, Aytemir K, Haznedaroğlu I, Ozer N, Ovünç K, Aksöyek S, Kes S, Kirazli S, Ozmen F. Increased plasma levels of soluble selectins in patients with unstable angina. Int J Cardiol. 2001;78(1):69–73.
74. Johnson-Tidey RR, McGregor JL, Taylor PR, Poston RN. Increase in the adhesion molecule P-selectin in endothelium overlying atherosclerotic plaques. Coexpression with intercellular adhesion molecule-1. Am J Pathol. 1994;144(5):952–61.
75. Preston RA, Coffey JO, Materson BJ, Ledford M, Alonso AB. Elevated platelet P-selectin expression and platelet activation in high risk patients with uncontrolled severe hypertension. Atherosclerosis. 2007;192(1):148–54.
76. Lip GY, Blann AD, Zarifis J, Beevers M, Lip PL, Beevers DG. Soluble adhesion molecule P-selectin and endothelial dysfunction in essential hypertension: implications for atherogenesis? A preliminary report. J Hypertens. 1995;13(12 Pt 2):1674–8.
77. Kling D, Stucki C, Kronenberg S, Tuerck D, Rhéaume E, Tardif JC, Gaudreault J, Schmitt C. Pharmacological control of platelet-leukocyte interactions by the human anti-P-selectin antibody inclacumab—preclinical and clinical studies. Thromb Res. 2013;131:401–10.
78. Halacheva K, Gulubova MV, Manolova I, Petkov D. Expression of ICAM-1, VCAM-1, E-selectin and TNF-alpha on the endothelium of femoral and iliac arteries in thromboangiitis obliterans. Acta Histochem. 2002;104(2):177–84.
79. Ley K. The role of selectins in inflammation and disease. Trends Mol Med. 2003;9(6):263–8.
80. Ramos CL, Huo Y, Jung U, Ghosh S, Manka DR, Sarembock IJ, Ley K. Direct demonstration of P-selectin- and VCAM-1-dependent mononuclear cell rolling in early atherosclerotic lesions of apolipoprotein E-deficient mice. Circ Res. 1999;84(11):1237–44.
81. Dong ZM, Chapman SM, Brown AA, Frenette PS, Hynes RO, Wagner DD. The combined role of P- and E-selectins in atherosclerosis. J Clin Invest. 1998;102(1):145–52.
82. Nadar SK, Al Yemeni E, Blann AD, Lip GY. Thrombomodulin, von Willebrand factor and E-selectin as plasma markers of endothelial damage/dysfunction and activation in pregnancy induced hypertension. Thromb Res. 2004;113(2):123–8.
83. Onozaki A, Sanada H, Midorikawa S, Imamura N, Hashimoto S, Watanabe T. E-selectin level may be a useful clinical marker for endothelial damage in hypertension. Am J Hypertens. 2000;13(S2):243A.

84. Freestone B, Chong AY, Nuttall S, Lip GY. Impaired flow mediated dilatation as evidence of endothelial dysfunction in chronic atrial fibrillation: relationship to plasma von Willebrand factor and soluble E-selectin levels. Thromb Res. 2008;122(1):85–90.

85. Ng CY, Kamisah Y, Faizah O, Jaarin K. Recycled deep-frying oil causes blood pressure elevation and vascular hypertrophy in Sprague–Dawley rats. Res Updates Med Sci. 2013; 1(1):2–6.

86. Porta B, Baldassarre D, Camera M, Amato M, Arquati M, Brusoni B, Fiorentini C, Montorsi P, Romano S, Tremoli E, Cortellaro M, MIAMI Study Group. E-selectin and TFPI are associated with carotid intima-media thickness in stable IHD patients: the baseline findings of the MIAMI study. Nutr Metab Cardiovasc Dis. 2008;18(4):320–8.

87. Schneider MP, Hilgers KF, Klingbeil AU, John S, Veelken R, Schmieder RE. Plasma endothelin is increased in early essential hypertension. Am J Hypertens. 2000;13(6 Pt 1): 579–85.

88. Douglas SA, Beck Jr GR, Elliott JD, Ohlstein EH. Pharmacologic evidence for the presence of three functional endothelin receptor subtypes in rabbit saphenous vein. J Cardiovasc Pharmacol. 1995;26 Suppl 3:S163–8.

89. Janakidevi K, Fisher MA, Del Vecchio PJ, Tiruppathi C, Figge J, Malik AB. Endothelin-1 stimulates DNA synthesis and proliferation of pulmonary artery smooth muscle cells. Am J Physiol. 1992;263(6 Pt 1):C1295–301.

90. Hynynen MM, Khalil RA. The vascular endothelin system in hypertension—recent patents and discoveries. Recent Pat Cardiovasc Drug Discov. 2006;1(1):95–108.

91. Schneider MP, Boesen EI, Pollock DM. Contrasting actions of endothelin ET(A) and ET(B) receptors in cardiovascular disease. Annu Rev Pharmacol Toxicol. 2007;47:731–59.

92. Leonard MG, Briyal S, Gulati A. Endothelin B receptor agonist, IRL-1620, provides long-term neuroprotection in cerebral ischemia in rats. Brain Res. 2012;1464:14–23.

93. Mazzuca MQ, Khalil RA. Vascular endothelin receptor type B: structure, function and dysregulation in vascular disease. Biochem Pharmacol. 2012;84(2):147–62.

94. Wilson JL, Taylor L, Polgar P. Endothelin-1 activation of ETB receptors leads to a reduced cellular proliferative rate and an increased cellular footprint. Exp Cell Res. 2012; 318(10):1125–33.

95. Stenman E, Malmsjö M, Uddman E, Gidö G, Wieloch T, Edvinsson L. Cerebral ischemia upregulates vascular endothelin ET(B) receptors in rat. Stroke. 2002;33(9):2311–6.

96. Zhang W, Li XJ, Zeng X, Shen DY, Liu CQ, Zhang HJ, Xu CB, Li XY. Activation of nuclear factor-κB pathway is responsible for tumor necrosis factor-α-induced up-regulation of endothelin B2 receptor expression in vascular smooth muscle cells in vitro. Toxicol Lett. 2012;209(2):107–12.

97. Maharaj AS, Saint-Geniez M, Maldonado AE, D'Amore PA. Vascular endothelial growth factor localization in the adult. Am J Pathol. 2006;168(2):639–48.

98. Ferrara N. Role of vascular endothelial growth factor in regulation of physiological angiogenesis. Am J Physiol Cell Physiol. 2001;280(6):C1358–66.

99. Ku DD, Zaleski JK, Liu S, Brock TA. Vascular endothelial growth factor induces EDRF-dependent relaxation in coronary arteries. Am J Physiol. 1993;265(2 Pt 2):H586–92.

100. Eskens FA, Verweij J. The clinical toxicity profile of vascular endothelial growth factor (VEGF) and vascular endothelial growth factor receptor (VEGFR) targeting angiogenesis inhibitors; a review. Eur J Cancer. 2006;42(18):3127–39.

101. Li B, Ogasawara AK, Yang R, Wei W, He GW, Zioncheck TF, Bunting S, de Vos AM, Jin H. KDR (VEGF receptor 2) is the major mediator for the hypotensive effect of VEGF. Hypertension. 2002;39(6):1095–100.

102. Husain K. Physical conditioning modulates rat cardiac vascular endothelial growth factor gene expression in nitric oxide-deficient hypertension. Biochem Biophys Res Commun. 2004;320(4):1169–74.

103. Bhargava P. VEGF kinase inhibitors: how do they cause hypertension? Am J Physiol Regul Integr Comp Physiol. 2009;297(1):R1–5.

104. Blann AD, Belgore FM, McCollum CN, Silverman S, Lip PL, Lip GY. Vascular endothelial growth factor and its receptor, Flt-1, in the plasma of patients with coronary or peripheral atherosclerosis, or Type II diabetes. Clin Sci. 2002;102(2):187–94.

105. Chung T, Kritharides L, Govindan M, Yiannikas J. Increased vascular endothelial growth factor (VEGF) expression in patients with pulmonary arterial hypertension. Heart Lung Circul. 2009;18 Suppl 3:S275.
106. Kojima T, Miyachi S, Sahara Y, Nakai K, Okamoto T, Hattori K, Kobayashi N, Hattori K, Negoro M, Yoshida J. The relationship between venous hypertension and expression of vascular endothelial growth factor: hemodynamic and immunohistochemical examinations in a rat venous hypertension model. Surg Neurol. 2007;68(3):277–84.
107. El Chami H, Hassoun PM. Immune and inflammatory mechanisms in pulmonary arterial hypertension. Prog Cardiovasc Dis. 2012;55(2):218–28.
108. Li X, Kumar A, Zhang F, Lee C, Tang Z. Complicated life, complicated VEGF-B. Trends Mol Med. 2012;18(2):119–27.
109. Franquni JV, do Nascimento AM, de Lima EM, Brasil GA, Heringer OA, Cassaro KO, da Cunha TV, Musso C, Silva Santos MC, Kalil IC, Endringer DC, Boëchat GA, Bissoli NS, de Andrade TU. Nandrolone decanoate determines cardiac remodelling and injury by an imbalance in cardiac inflammatory cytokines and ACE activity, blunting of the Bezold-Jarisch reflex, resulting in the development of hypertension. Steroids. 2013;78(3):379–85.
110. Li XQ, Cao W, Li T, Zeng AG, Hao LL, Zhang XN, Mei QB. Amlodipine inhibits TNF-alpha production and attenuates cardiac dysfunction induced by lipopolysaccharide involving PI3K/Akt pathway. Int Immunopharmacol. 2009;9(9):1032–41.
111. Nepal S, Malik S, Sharma AK, Bharti S, Kumar N, Siddiqui KM, Bhatia J, Kumari S, Arya DS. Abresham ameliorates dyslipidemia, hepatic steatosis and hypertension in high-fat diet fed rats by repressing oxidative stress, TNF-α and normalizing NO production. Exp Toxicol Pathol. 2012;64(7–8):705–12.
112. Wang Q, Zuo XR, Wang YY, Xie WP, Wang H, Zhang M. Monocrotaline-induced pulmonary arterial hypertension is attenuated by TNF-α antagonists via the suppression of TNF-α expression and NF-κB pathway in rats. Vascul Pharmacol. 2013;58(1–2):71–7.
113. Paz Y, Frolkis I, Pevni D, Shapira I, Yuhas Y, Iaina A, Wollman Y, Chernichovski T, Nesher N, Locker C, Mohr R, Uretzky G. Effect of tumor necrosis factor-alpha on endothelial and inducible nitric oxide synthase messenger ribonucleic acid expression and nitric oxide synthesis in ischemic and nonischemic isolated rat heart. J Am Coll Cardiol. 2003;42(7):1299–305.
114. Chia S, Qadan M, Newton R, Ludlam CA, Fox KA, Newby DE. Intra-arterial tumor necrosis factor-alpha impairs endothelium-dependent vasodilatation and stimulates local tissue plasminogen activator release in humans. Arterioscler Thromb Vasc Biol. 2003;23(4):695–701.
115. Förstermann U, Sessa WC. Nitric oxide synthases: regulation and function. Eur Heart J. 2012;33(7):829–37.
116. Kuo PC, Schroeder RA. The emerging multifaceted roles of nitric oxide. Ann Surg. 1995;221(3):220–35.
117. Morris Jr SM, Billiar TR. New insights into the regulation of inducible nitric oxide synthesis. Am J Physiol. 1994;266(6 Pt 1):E829–39.
118. Perrotta I, Brunelli E, Sciangula A, Conforti F, Perrotta E, Tripepi S, Donato G, Cassese M. iNOS induction and PARP-1 activation in human atherosclerotic lesions: an immunohistochemical and ultrastructural approach. Cardiovasc Pathol. 2011;20(4):195–203.
119. Castro MM, Rizzi E, Rodrigues GJ, Ceron CS, Bendhack LM, Gerlach RF, Tanus-Santos JE. Antioxidant treatment reduces matrix metalloproteinase-2-induced vascular changes in renovascular hypertension. Free Radic Biol Med. 2009;46(9):1298–307.
120. Vaziri ND, Rodríguez-Iturbe B. Mechanisms of disease: oxidative stress and inflammation in the pathogenesis of hypertension. Nat Clin Pract Nephrol. 2006;2(10):582–93.
121. Huang CF, Hsu CN, Chien SJ, Lin YJ, Huang LT, Tain YL. Aminoguanidine attenuates hypertension, whereas 7-nitroindazole exacerbates kidney damage in spontaneously hypertensive rats: the role of nitric oxide. Eur J Pharmacol. 2013;699(1–3):233–40.
122. Soskić SS, Dobutović BD, Sudar EM, Obradović MM, Nikolić DM, Djordjevic JD, Radak DJ, Mikhailidis DP, Isenović ER. Regulation of inducible nitric oxide synthase (iNOS) and its potential role in insulin resistance, diabetes and heart failure. Open Cardiovasc Med J. 2011;5:153–63.

123. Su NY, Tsai PS, Huang CJ. Clonidine-induced enhancement of iNOS expression involves NF-kappaB. J Surg Res. 2008;149(1):131–7.
124. Oliveira-Paula GH, Lacchini R, Coeli-Lacchini FB, Junior HM, Tanus-Santos JE. Inducible nitric oxide synthase haplotype associated with hypertension and responsiveness to antihypertensive drug therapy. Gene. 2013;515(2):391–5.
125. Bader M, Peters J, Baltatu O, Müller DN, Luft FC, Ganten D. Tissue renin-angiotensin systems: new insights from experimental animal models in hypertension research. J Mol Med. 2001;79(2–3):76–102.
126. Deji N, Kume S, Araki S, Isshiki K, Araki H, Chin-Kanasaki M, Tanaka Y, Nishiyama A, Koya D, Haneda M, Kashiwagi A, Maegawa H, Uzu T. Role of angiotensin II-mediated AMPK inactivation on obesity-related salt-sensitive hypertension. Biochem Biophys Res Commun. 2012;418(3):559–64.
127. Sakamoto A, Hongo M, Furuta K, Saito K, Nagai R, Ishizaka N. Pioglitazone ameliorates systolic and diastolic cardiac dysfunction in rat model of angiotensin II-induced hypertension. Int J Cardiol. 2012;167:409–15.
128. Xu S, Zhi H, Hou X, Jiang B. Angiotensin II modulates interleukin-1β-induced inflammatory gene expression in vascular smooth muscle cells via interfering with ERK-NF-κB crosstalk. Biochem Biophys Res Commun. 2011;410(3):543–8.
129. Zhu N, Zhang D, Chen S, Liu X, Lin L, Huang X, Guo Z, Liu J, Wang Y, Yuan W, Qin Y. Endothelial enriched microRNAs regulate angiotensin II-induced endothelial inflammation and migration. Atherosclerosis. 2011;215(2):286–93.
130. Schieffer B, Bünte C, Witte J, Hoeper K, Böger RH, Schwedhelm E, Drexler H. Comparative effects of AT1-antagonism and angiotensin-converting enzyme inhibition on markers of inflammation and platelet aggregation in patients with coronary artery disease. J Am Coll Cardiol. 2004;44(2):362–8.
131. Schmeisser A, Soehnlein O, Illmer T, Lorenz HM, Eskafi S, Roerick O, Gabler C, Strasser R, Daniel WG, Garlichs CD. ACE inhibition lowers angiotensin II-induced chemokine expression by reduction of NF-kappaB activity and AT1 receptor expression. Biochem Biophys Res Commun. 2004;325(2):532–40.
132. Tsuneki H, Tokai E, Suzuki T, Seki T, Okubo K, Wada T, Okamoto T, Koya S, Kimura I, Sasaoka T. Protective effects of coenzyme Q10 against angiotensin II-induced oxidative stress in human umbilical vein endothelial cells. Eur J Pharmacol. 2013;701(1–3):218–27.
133. Kamper M, Tsimpoukidi O, Chatzigeorgiou A, Lymberi M, Kamper EF. The antioxidant effect of angiotensin II receptor blocker, losartan, in streptozotocin-induced diabetic rats. Transl Res. 2010;156(1):26–36.
134. Khaper N, Singal PK. Modulation of oxidative stress by a selective inhibition of angiotensin II type 1 receptors in MI rats. J Am Coll Cardiol. 2001;37(5):1461–6.
135. Oudot A, Vergely C, Ecarnot-Laubriet A, Rochette L. Angiotensin II activates NADPH oxidase in isolated rat hearts subjected to ischaemia-reperfusion. Eur J Pharmacol. 2003;462(1–3):145–54.
136. Crowley SD, Gurley SB, Herrera MJ, Ruiz P, Griffiths R, Kumar AP, Kim HS, Smithies O, Le TH, Coffman TM. Angiotensin II causes hypertension and cardiac hypertrophy through its receptors in the kidney. Proc Natl Acad Sci USA. 2006;103(47):17985–90.
137. Theodosiou Z, Kasampalidis IN, Livanos G, Zervakis M, Pitas I, Lyroudia K. Automated analysis of FISH and immunohistochemistry images: a review. Cytometry A. 2007;71(7):439–50.
138. Hlawaty H, Suffee N, Sutton A, Oudar O, Haddad O, Ollivier V, Laguillier-Morizot C, Gattegno L, Letourneur D, Charnaux N. Low molecular weight fucoidan prevents intimal hyperplasia in rat injured thoracic aorta through the modulation of matrix metalloproteinase-2 expression. Biochem Pharmacol. 2011;81(2):233–43.
139. Shang T, Liu Z, Zhou M, Zarins CK, Xu C, Liu CJ. Inhibition of experimental abdominal aortic aneurysm in a rat model by way of tanshinone IIA. J Surg Res. 2012;178(2):1029–37.
140. Wang LJ, Yu YH, Zhang LG, Wang Y, Niu N, Li Q, Guo LM. Taurine rescues vascular endothelial dysfunction in streptozocin-induced diabetic rats: correlated with downregulation of LOX-1 and ICAM-1 expression on aortas. Eur J Pharmacol. 2008;597(1–3):75–80.

141. Martínez AC, Stankevicius E, Jakobsen P, Simonsen U. Blunted non-nitric oxide vasodilatory neurotransmission in penile arteries from renal hypertensive rats. Vascul Pharmacol. 2006;44(5):354–62.
142. Bölükbaşı Hatip FF, Hatip-Al-Khatib I. Effects of β-sheet breaker peptides on altered responses of thoracic aorta in rats' Alzheimer's disease model induced by intraamygdaloid Aβ40. Life Sci. 2013;92(3):228–36.
143. D'Amico F, Skarmoutsou E, Stivala F. State of the art in antigen retrieval for immunohisto-chemistry. J Immunol Methods. 2009;341(1–2):1–18.
144. Werner M, Chott A, Fabiano A, Battifora H. Effect of formalin tissue fixation and processing on immunohistochemistry. Am J Surg Pathol. 2000;24(7):1016–9.
145. De Marzo AM, Fedor HH, Gage WR, Rubin MA. Inadequate formalin fixation decreases reliability of p27 immunohistochemical staining: probing optimal fixation time using high-density tissue microarrays. Hum Pathol. 2002;33(7):756–60.
146. Takai H, Kato A, Ishiguro T, Kinoshita Y, Karasawa Y, Otani Y, Sugimoto M, Suzuki M, Kataoka H. Optimization of tissue processing for immunohistochemistry for the detection of human glypican-3. Acta Histochem. 2010;112(3):240–50.
147. Varma M. Immunohistochemistry in prostate neoplasia: pitfalls and progress. Diagn Histopathol. 2011;17(10):447–53.
148. Larsson LI. Immunocytochemistry: theory and practice. Boca Raton, FL: CRC Press; 1988.
149. Huang SN, Minassian H, More JD. Application of immunofluorescent staining on paraffin sections improved by trypsin digestion. Lab Invest. 1976;35(4):383–90.
150. Yamashita S. Heat-induced antigen retrieval: mechanisms and application to histochemistry. Prog Histochem Cytochem. 2007;41(3):141–200.
151. Gown AM, de Wever N, Battifora H. Microwave-based antigenic unmasking. A revolution-ary new technique for routine immunohistochemistry. Appl Immunohistochem Mol Morphol. 1993;1:256–66.
152. Volkin DB, Klibanov AM. Minimizing protein inactivation. In: Creighton TE, editor. Protein function: a practical approach. Oxford: IRL Press; 1989. p. 1–24.
153. Kim SH, Kook MC, Shin YK, Park SH, Song HG. Evaluation of antigen retrieval buffer systems. J Mol Histol. 2004;35(4):409–16.
154. Sukhotnik I, Greenblatt R, Voskoboinik K, Lurie M, Coran AG, Mogilner JG. Relationship between time of reperfusion and E-selectin expression, neutrophil recruitment, and germ cell apoptosis after testicular ischemia in a rat model. Fertil Steril. 2008;90(4 Suppl):1517–22.
155. Ishida H, Kogaki S, Ichimori H, Narita J, Nawa N, Ueno T, Takahashi K, Kayatani F, Kishimoto H, Nakayama M, Sawa Y, Beghetti M, Ozono K. Overexpression of endothelin-1 and endothelin receptors in the pulmonary arteries of failed Fontan patients. Int J Cardiol. 2012;159(1):34–9.
156. Nobre MEP, Correia AO, de Brito Borges M, Sampaio TMA, Chakraborty SA, de Oliveira Gonçalves D, de Castro Brito GA, Leal LKAM, Felipe CFB, Lucetti DL, Arida RM, de Barros Viana GS. Eicosapentaenoic acid and docosahexaenoic acid exert anti-inflammatory and antinociceptive effects in rodents at low doses. Nutr Res. 2013;33:422–33.
157. Zaitone SA, Abo-Gresha NM. Rosuvastatin promotes angiogenesis and reverses isoproterenol-induced acute myocardial infarction in rats: role of iNOS and VEGF. Eur J Pharmacol. 2012;691(1–3):134–42.
158. Harata T, Ando H, Iwase A, Nagasaka T, Mizutani S, Kikkawa F. Localization of angiotensin II, the AT1 receptor, angiotensin-converting enzyme, aminopeptidase A, adipocyte-derived leucine aminopeptidase, and vascular endothelial growth factor in the human ovary through-out the menstrual cycle. Fertil Steril. 2006;86(2):433–9.
159. Thomas MA, Lemmer B. The use of heat-induced hydrolysis in immunohistochemistry on angiotensin II (AT1) receptors enhances the immunoreactivity in paraformaldehyde-fixed brain tissue of normotensive Sprague–Dawley rats. Brain Res. 2006;1119(1):150–64.
160. Chu PG, Chang KL, Arber DA, Weiss LM. Practical applications of immunohistochemistry in hematolymphoid neoplasms. Ann Diagn Pathol. 1999;3(2):104–33.

161. Renshaw S. Immunochemical staining techniques. In: Renshaw S, editor. Immunohistochemistry: methods express. Bloxham: Scion Publishing Ltd; 2007. p. 45–96.
162. Koga T, Kwan P, Zubik L, Ameho C, Smith D, Meydani M. Vitamin E supplementation suppresses macrophage accumulation and endothelial cell expression of adhesion molecules in the aorta of hypercholesterolemic rabbits. Atherosclerosis. 2004;176(2):265–72.
163. Faratzis G, Tsiambas E, Rapidis AD, Machaira A, Xiromeritis K, Patsouris E. VEGF and ki 67 expression in squamous cell carcinoma of the tongue: an immunohistochemical and computerized image analysis study. Oral Oncol. 2009;45(7):584–8.
164. Kaczmarek E, Górna A, Majewski P. Techniques of image analysis for quantitative immunohistochemistry. Annales Academiae Medicae Bialostocensis. 2004;49(1):155–8.
165. Kokolakis G, Panagis L, Stathopoulos E, Giannikaki E, Tosca A, Krüger-Krasagakis S. From the protein to the graph: how to quantify immunohistochemistry staining of the skin using digital imaging. J Immunol Methods. 2008;331(1–2):140–6.
166. Lang DS, Heilenkötter U, Schumm W, Behrens O, Simon R, Vollmer E, Goldmann T. Optimized immunohistochemistry in combination with image analysis: a reliable alternative to quantitative ELISA determination of uPA and PAI-1 for routine risk group discrimination in breast cancer. Breast. 2013;22:736–43.
167. Matkowskyj KA, Schonfeld D, Benya RV. Quantitative immunohistochemistry by measuring cumulative signal strength using commercially available software photoshop and matlab. J Histochem Cytochem. 2000;48(2):303–12.
168. Samsi S, Krishnamurthy AK, Gurcan MN. An efficient computational framework for the analysis of whole slide images: application to follicular lymphoma immunohistochemistry. J Comput Sci. 2012;3(5):269–79.
169. Di Cataldo S, Ficarra E, Acquaviva A, Macii E. Automated segmentation of tissue images for computerized IHC analysis. Comput Methods Programs Biomed. 2010;100(1):1–15.
170. Di Cataldo S, Ficarra E, Macii E. Computer-aided techniques for chromogenic immunohistochemistry: status and directions. Comput Biol Med. 2012;42(10):1012–25.

Chapter 9
Role of Lymphoid/Mucosa-Associated Lymphoid Tissue Markers in Toxicological Immunohistochemistry

Kuldeep Singh, E. Driskell, and A. Stern

Introduction

The lymphatic system is a vascular system that runs parallel to the blood vascular system and drains interstitial fluid that is mostly generated from the blood. Interstitial fluid is drained to the regional lymph nodes and finally to the venous system as lymph. Lymph nodes, the spleen, thymus, and mucosa-associated lymphoid tissue (MALT) are part of the lymphatic system and are divided into primary and secondary lymphoid organs. Primary lymphoid organs are sites at which B and T lymphocytes of the immune system are formed and mature in an antigen-independent fashion and include the thymus and bone marrow. Secondary lymphoid organs are the sites where mature lymphocytes "home" after exiting primary lymphoid organs and undergo selection and clonal expansion in an antigen-dependent fashion and include the spleen, lymph nodes, and MALT.

K. Singh, DVM, MS, PhD, Diplomate ACVP (✉)
Department of Pathobiology, University of Illinois, Urbana-Champaign, Urbana, IL, USA

Veterinary Diagnostic Laboratory, University of Illinois, Urbana-Champaign, 1224 Veterinary Basic Science Building, 2001 South Lincoln, Urbana, IL 61802, USA
e-mail: ksingh08@illinois.edu

E. Driskell, DVM, PhD, Diplomate ACVP
Department of Pathobiology, University of Illinois, Urbana, IL 61802, USA

Veterinary Diagnostic Laboratory, University of Illinois, Urbana, IL 61802, USA

A. Stern, DVM, CMI-IV, CFC, Diplomate ACVP
Veterinary Diagnostic Laboratory, University of Illinois, Urbana-Champaign, Urbana, IL, USA

© Springer Science+Business Media New York 2016
S.A. Aziz, R. Mehta (eds.), *Technical Aspects of Toxicological Immunohistochemistry*, DOI 10.1007/978-1-4939-1516-3_9

Lymph Node

Lymph nodes are integral part of the lymphatic system and are encapsulated by a connective tissue capsule. The parenchyma of lymph nodes is organized into cortex, paracortex, and medulla which are supported by fibrous trabeculae and a reticulum of fibroblastic reticular cells. The cortex consists of lymphoid follicles (B cell centers) whereas the paracortex contains sheets of T cells (Fig. 9.1). The medulla is divided into medullary cords and medullary sinuses. Medullary cords are populated by plasma cells, lymphocytes, and macrophages, and medullary sinuses contain stellate reticular cells and are often populated by macrophages and antigen-presenting dendritic cells which phagocytize various antigens and nonantigenic particles, e.g., carbon. The reticular cells have an immunologic function of guiding lymphocytes and antigen-presenting cells to facilitate their interaction with B and T lymphocytes.

Lymph nodes filter interstitial fluid which is drained by the afferent lymphatic vessels into the subcapsular sinuses of the regional lymph nodes. Lymph filtration helps in removal of foreign substances including infectious agents (bacteria, viruses, and parasites) and facilitates surveillance of incoming antigens and their interaction with B and T lymphocytes. Lymph nodes also play a critical role in the humoral and cell-mediated immune response by producing plasma cells which produce antibodies and T cells which contribute to both humoral and cell-mediated immune response.

The cortex contains lymphoid follicles which are morphologically and physiologically similar to that of splenic or MALT lymphoid follicles. There are two types of lymphoid follicles—primary and secondary. Mature naive B lymphocytes expressing specific receptor for specific antigens exit primary lymphoid organs and circulate the blood and lymphatics and home to secondary lymphoid organs [1]. On their arrival in secondary lymphoid organs, B lymphocytes exit through high endothelial venules (HEV) and home to the primary follicles. Those B lymphocytes that recognize the antigen undergo clonal expansion to form secondary lymphoid follicles characterized histologically by prominent germinal centers. Plasma cell precursors formed in germinal centers migrate to the medullary cords and produce monoclonal antibodies against specific antigen (Fig. 9.1) [1]. Sinuses contain veins and arteries and may be surrounded by macrophages. The gaps in the lining of medullary sinuses permit passage of cell from the efferent lymph directly to the medullary areas.

Hyperplastic Lesions

Hyperplasia in a lymph node can involve cortex (B cells), paracortex (T cells), medullary chords (plasma cells), and macrophages. Hyperplastic lesions of B cells are characterized by formation of lymphoid follicles with a prominent germinal

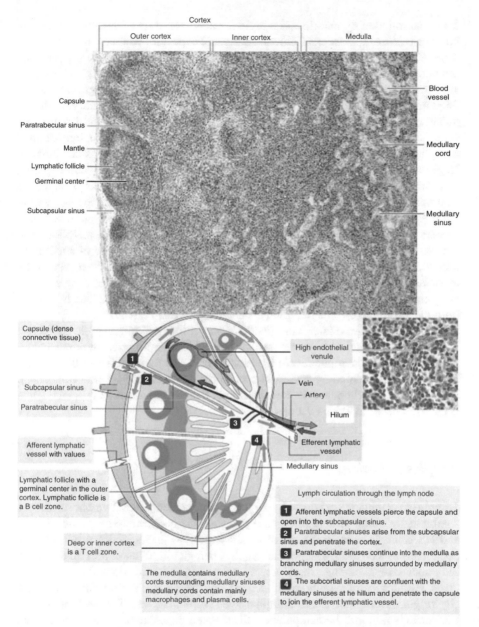

Fig. 9.1 Kierszenbaum AL. Histology and cell biology; an introduction to pathology. Immune-lymphatic system. p. 307, Mosby Elsevier, 2007. Reproduced with permission from Mosby Elsevier publication. Copyright © 2012, 2007, 2002 by Saunders, an imprint of Elsevier, Inc

center surrounded by mantle and marginal zones. The time to develop follicles in humans is closer to 2 weeks following antigenic exposure, whereas in dogs it is 10 days. Germinal centers are present predominantly in the cortex but may populate the paracortex with sustained antigenic stimulation. Germinal centers have a consistent architecture with the interior cells having a superficial or light pole and a deep or dark pole. Recognition of this polarity is important observation in distinguishing lymphoid hyperplasia from follicular lymphoma [2]. The cells within the superficial pole are small lymphocytes with a moderate amount of light-stained cytoplasm giving it a less dense appearance, whereas the cells of the deep pole are large lymphocytes with proportionally less cytoplasm that is deeply amphophilic, resulting in darker and compact appearance. Many tingible body macrophages are also scattered within the germinal centers and phagocytize lymphocytes whose antibody exhibits too high or too low avidity. Generally, well-developed germinal centers are not present at birth unless there has been intrauterine infection. The mantle cell zone is composed of relatively homogenous population of small lymphocytes that have round compact nuclei that lack nucleoli and have very little cytoplasm and give rise to pregerminal center B cells [2]. Marginal zone cells are not generally prominent in normal nodes but may become prominent in case of immune response [2]. The marginal zone cells have relatively abundant cytoplasm and round and larger nuclei with fine dispersed chromatin and a single nucleolus. These cells are believed to be memory cells [2]. Medullary cords are populated by plasma cells and macrophages. In the case of increased demand for erythropoiesis, cords can be populated by erythroid progenitors (extramedullary hematopoiesis). The immunohistochemical markers and their sources for macrophages and T and B cells are listed in Tables 9.1 and 9.2.

Neoplastic Lesions

Tumors of lymphocyte origin are lymphosarcoma and leukemia. Lymphosarcoma is a solid tissue tumor arising in organs other than the bone marrow and are of B, T, and NK cell origin. However, in advanced stages, neoplastic cell can infiltrate the bone marrow. The term "lymphoma" implies benign tumor but almost all advanced lymphomas are malignant; therefore, "lymphosarcoma" is considered a better diagnosis. Histologically, lymphosarcoma can be diffuse or of nodular/follicular hierarchy origin, i.e., mantle zone, marginal zone, follicular, and T zone lymphomas. Leukemia is tumor of hematopoietic cells that originates in the bone marrow, e.g., RBC, neutrophils, eosinophils, megakaryocytes, lymphocytes, and monocytes. Neoplastic cells are present in the blood and/or bone marrow.

Lymphosarcoma in genetically intact mice often arises in the mesenteric lymph nodes, spleen, and intestinal Peyer's patches. Involvement of lymph nodes can be primary or secondary [3]. In case of primary involvement, tumors originate in the lymph nodes, whereas they metastasize to regional lymph nodes in secondary

Table 9.1 Immunohistochemical markers of the immune system on paraffin sections in mouse

Antibody	Major cells expressing antigen	Source	Product number
Bcl-2	Germinal center B cells	Santa Cruz Biotechnology	Sc-7382
Bcl-6	Germinal center B cells	Santa Cruz Biotechnology	Sc-858
Caspase-3 cleaved	Apoptotic cells	Cell Signaling Technology	2H12
CD3	T lymphocytes	Dako	A0452
CD19	B cells	Cell Signaling Technology	3574
CD21	B cells, follicular dendritic cells	Santa Cruz Biotechnology	Sc-7028FITC
CD30	Activated T and B cells	BD Biosciences	553824
CD43	T and B cells, myeloid cells	BD Biosciences	552366
CD45R/B220	B cells	BD Biosciences	553850
CD138 (Syndecan-1)	B cells, plasma cells	BD Biosciences	553714
F4/80	Histiocytes	Serotec	MCA497BB
IRF4	B cells, plasma cells	Santa Cruz Biotechnology	Sc-6059
Mac-2	Histiocytes	Cedarlane Laboratories	CL8942AP
Myeloperoxidase	Myeloid cells	Dako	A0398
Nitric oxide synthase (NOS2)	Myeloid cells	Santa Cruz Biotechnology	Sc-651
Pax5	B cells	Santa Cruz Biotechnology	Sc-1974
PNA-biotin	B cells, follicles, germinal centers	Vector Laboratories	B-1075
PU.1	B cells	Santa Cruz Biotechnology	Sc-5949
TdT	T cells	Lab Vision	Ab-2

Source: Ward J, et al. Immunohistochemical markers for the rodent immune system. Toxicol Pathol. 2006;34:616–30

involvement. The origin of metastases depends largely on the location of the primary neoplasm and subsequent spread to the regional lymph node, e.g., cervical and peritracheal lymph nodes are the most common metastatic sites for thyroid carcinomas. The mesenteric lymph nodes are common site of metastasis for small and large enteric carcinomas and enteric lymphoma. The mandibular lymph node is a common site for malignant Zymbal gland carcinoma metastasis.

Table 9.2 Immunohistochemical markers of the immune system on paraffin sections in rat

Antibody	Major cells expressing antigen	Source	Product number
CD3	T lymphocytes	Dako	A0452
CD4	T-helper cells	Serotec	MCA55G
CD5	Subset of T lymphocytes, NK cells	Serotec	MCA52G
CD8	Subset of T cells, NK cells	Serotec	MCA48G
CD25	Activated T cells	Serotec	MCA273R
CD43	All leukocytes excluding B cells	Serotec	MCA54G
CD45RC	B cells	Serotec	MCA53R
CD68	Myeloid cells	Serotec	MCA341R
IgM	B cells	Vector	BA-2020
MHC class II	B cells	Serotec	MCA46G
Immunoglobulin human kappa light chains	Immunoglobulin-producing B cells	Dako	A0191

Source: Ward J, et al. Immunohistochemical markers for the rodent immune system. Toxicol Pathol. 2006;34:616–30

Toxicologic Lesions

The evaluation of a potential treatment-related effect and subsequent alteration in the gross and histologic morphology depend on the physical and chemical properties of toxicologic agent or test compound, site of administration, and duration of the study. For example, oral administration of a paraffin compound resulted in granulomatous inflammation in the mesenteric lymph nodes with accumulation of macrophages in sinuses [4]. However, common toxicologic lesions include lymphocyte degeneration, necrosis, apoptosis, depletion, atrophy, and lymphoid and macrophage hyperplasia. A survey of nonneoplastic lesions in the mandibular and mesenteric lymph nodes in control B6C3F1 mice revealed lymphoid hyperplasia, plasma cell hyperplasia, lymphoid depletion, inflammation, sinus histiocytosis, angiectasis, hemorrhage, pigmentation, and extramedullary hematopoiesis [3].

Inflammation

Lymph node inflammation (lymphadenitis) can be acute or chronic. Lymph nodes in acute lymphadenitis are grossly enlarged and swollen and may be red. On cut section, the capsule is tight and the cut parenchyma is dark red and wet. Microscopically, it is characterized by an exudative lesion accompanied by hemorrhages and neutrophils with or without necrosis. Chronic lymphadenitis is characterized by grossly enlarged, firm to soft, pale, and non-exudative morphology [1]. On cut section, the parenchyma is dry and pale. Microscopically, these lymph nodes contain mostly mononuclear cells (histiocytes) which are accompanied by fibrosis and plasma cell and lymphoid hyperplasia. Because lymph nodes drain local region, lymphadenitis is commonly observed as a secondary lesion due to local drainage of an adjacent primary inflammatory or neoplastic lesion.

Angiectasis

Angiectasis is common in mesenteric lymph nodes of some mouse strains and is characterized by dilation of veins in the medulla, capsule, and in the surrounding adipose tissue. It is often accompanied by hemorrhages [3]. These lesions must be differentiated from tumors of endothelial origin—hemangioma and hemangiosarcoma. Distinction between hemangioma and angiectasis can be difficult. However, hemangiosarcoma is characterized by formation of cavernous channels lined by pleomorphic neoplasia with variation in shape and size of cytoplasm and nuclei and increased mitotic index.

Degenerative Lesions

1. Atrophy results into decreased size of lymph nodes and is seen in older animals, stress, radiation exposure, and chemotherapeutic agents and corticosteroid treatment. Grossly, these lymph nodes are small and contain indistinct cortex and medulla. Microscopically, there is reduced number of cortical, paracortical, and medullary lymphocytes and lack lymphoid follicles [3]. There may be fatty tissue infiltration and collagen deposition [5].
2. Medullary pigmentation including hemosiderin (iron), carbon, and other pigments can sometimes be present in macrophages in the medulla.
3. Amyloid deposition is found in subcapsular sinuses and paracortex.

Gut-Associated Lymphoid Tissue

Similar to other mucosa-associated lymphoid tissues such as BALT and NALT, gut-associated lymphoid tissues (GALTs) are dispersed aggregates of nonencapsulated organized lymphoid tissue throughout the gastrointestinal tract and are associated with localized immune responses at mucosal surfaces. Gut-associated lymphoid tissue is the site of intense B cell activation, proliferation, and terminal differentiation in response to foreign antigen, particularly, bacteria, and there is production of large amounts of IgA within the GALT [6].

Within the small intestine, several types of lymphoid tissue have been described and include: intraepithelial lymphocytes, Peyer's patches, isolated lymphoid follicles,

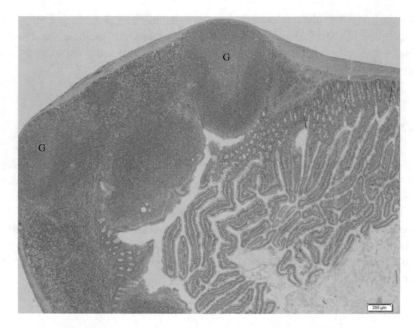

Fig. 9.2 The small intestine, Peyer's patch, mouse. Peyer's patches are randomly distributed within the small intestine with greatest intensity within the jejunum. Germinal center (G). Hematoxylin and eosin, ×4. *Bar* = 200 µm

and cryptopatches [7, 8]. Lymphoglandular complexes have been described within the large intestine [8]. Peyer's patches and lymphoglandular complexes are primary induction sites within the small and large intestine, respectively [8]. Mice deficient in the IL-7 receptor alpha chain (Il-7r alpha −/−) have disruption in the formation of Peyer's patches [9].

Peyer's Patches

Peyer's patches, similar to other peripheral lymphoid tissues, consist of lymphocytes, macrophages, and supporting stromal components (i.e., dendritic cells) [10]. Peyer's patches have a random distribution within the mucosa and submucosa with an antimesenteric orientation within the small intestine. The jejunum has the greatest density of Peyer's patches (Fig. 9.2) [8]. Peyer's patches are covered by follicle-associated epithelium; M cells, a component of follicle-associated epithelium, uptake antigens from within the intestinal lumen and pass them to antigen-presenting cells. Approximately 6–12 basally located germinal centers are found within the Peyer's patch of rodents [8]. The follicular aggregates are composed of B lymphocytes and the interfollicular areas are composed of T lymphocytes. The B cell area is larger than the T cell area and the T cell to B cell ratio is 0.2 [11]. In the Fischer 344 rat, the majority of lymphocytes in Peyer's patches are B lymphocytes (56 %) [12].

The size, number, distribution, and composition of Peyer's patches can vary depending on species or strain. Fischer 344 rats have smaller Peyer's patches than Wistar rats [13]. Lewis rats have nearly equal ratios of CD4+ and CD8+ T cells, and Fischer 344 rats have a 5.0 ratio of CD4+ to CD8+ T cells [8, 11]. Fischer 344 rat Peyer's patch numbers do not change as a result of aging; however, one study suggests that the CD8+ cell distribution at inductor and effector sites of GALT undergoes age-related shifts [14]. In the Fischer 344 rat, the majority of lymphocytes in Peyer's patches are B lymphocytes (56 %) [12]. Cigarette smoke exposure increases apoptosis in the follicle-associated epithelium in Peyer's patches of rats [15]. Granulomatous inflammation and scarring in Peyer's patches have been reported following exposure to high fatty acid diets in Wistar rats [16].

Lymphoid Follicles

Isolated lymphoid follicles are found within the small intestine in an antimesenteric location and can be found in the rat, mouse, rabbit, and guinea pig. They are smaller than the Peyer's patch, have a barrel-shaped lymphoid aggregate, and are found within shorter villi. There are 1–2 B cell follicles that may contain germinal centers [8]. The overlying follicle-associated epithelium contains M cells, scattered dendritic cells, and fewer macrophages.

Cryptopatches

Cryptopatches are found within the intercryptal lamina propria and contain aggregates of T cells and dendritic cells. The diameter of the cryptopatches is approximately 80 μm [8]. Formation of cryptopatches occurs after weaning and there are up to 1700 per adult mouse [8].

Lymphoglandular Complexes

Lymphoglandular complexes are found within the colon and resemble Peyer's patches. They contain smaller and fewer follicles with small germinal centers. In the mouse there is a random distribution of lymphoglandular complexes within the distal colon. In the mouse, there are approximately 1.4 lymphoglandular complexes per centimeter of colon, and the rectal patch (lymphoglandular complex) is found 10 mm proximal to the anus [8].

Various studies have utilized immunohistochemical markers to characterize GALT in mice and rats within formalin-fixed paraffin-embedded tissues. Both mouse and rat T cells are labeled with CD3, and mouse B cells are consistently labeled with CD45R/B220 [17]. Immunohistochemical labeling of T cells for CD4 and CD8 can be performed on formalin-fixed tissues in the rat [14, 17]. Caspase-3 cleaved has been reported to be a marker for apoptotic cells in the mouse [17].

Bronchus-Associated Lymphoid Tissue

Bronchus-associated lymphoid tissue (BALT) is composed of lymphoid aggregates present in the bronchial submucosa. The distribution of BALT along the airways is random but is most often located at the bifurcation of bronchi and bronchioles and is always located between a bronchus and an artery [8, 12]. BALT is present in both the rat and the mouse; however, the prominence of the BALT differs between the two species. Additionally, BALT prominence can also differ between strains of mouse, emphasizing the need for comparative controls [18]. No special sectioning of the lung is required to examine the BALT.

In the rat, BALT is organized into B cell zones and T cell zones, but these zones are not consistent between BALT nodules [18]. The T cells may be confined to discrete zones with the B cells present in collections adjacent to the bronchial epithelium and a scattering of T cells within the B cell zones. Germinal centers do not normally occur in BALT but can be induced by treatment [18]. Several studies have demonstrated the T cell to B cell ratio in rat BALT is approximately 2:3, although the area occupied by T cells versus that of the B cells in nodules of BALT is approximately equivalent [8, 12, 18]. T-helper cells are the dominant population of T cells in the rat BALT, with a reported ratio of 2:1 for T-helper lymphocytes: cytotoxic T lymphocytes [18]. The BALT of the mouse is organized much like that of the rat; however, BALT in mice is poorly developed compared to the rat and is described only in certain strains [12]. In fact, the presence of BALT in germ-free mice is debated [8].

The BALT lies under regions of follicle-associated epithelium of the adjacent airway. This epithelium is attenuated and contains microfold (M) cells but no goblet or ciliated cells [8]. Interfollicular areas of the BALT can be identified by the presence of high endothelial venules. The BALT contains cells other than lymphocytes, including fibroblasts, reticulum cells, macrophages, and follicular and intrafollicular dendritic cells, and there is a higher proportion of collagen and reticular fibers compared to lymph nodes [12].

Hyperplastic Lesions

Lymphocytes within the BALT should not infiltrate the basal lamina in normal conditions, but this can be induced by the administration of intratracheal antigens [12]. BALT will also exhibit lymphocyte expansion with intratracheal antigen administration that stimulates T cells [18]. Macrophage aggregates can occasionally be observed in the BALT due to uptake of specific substances or infectious organisms [16].

Immunohistochemical Markers for BALT

Various studies have used immunohistochemical markers to characterize BALT; however, consistency of these markers has not been extensively studied in BALT. Murine B cells are reported to be consistently labeled with B220 (CD45R/B220) [17, 19]. A consistent pan T cell marker in mice is CD3 [19]. Differentiation of T cells can be done with CD4 and CD8 markers; however, current antibodies for CD4 and CD8 do not work well in formalin-fixed tissues, and therefore, CD4 and CD8 cells must be identified by flow cytometry or immunohistochemistry of frozen section [19]. CD 138 labels plasma cells and Mac-2, Mac-3, and lysozyme are consistent markers for macrophages [17, 19]. Caspase-3 cleaved has been reported to work as an apoptosis marker in mice [17].

Nasal-Associated Lymphoid Tissue

In mice and rats, nasopharynx-associated lymphoid tissue (NALT) is the equivalent of the Waldeyer ring in other species. The Waldeyer ring is composed of lymphoid tissue in the nasopharynx, oropharynx, and laryngopharynx (tonsils). The mouse and the rat only have well-developed lymphoid tissue in the nasopharynx; therefore, the term "NALT" rather than tonsil is used to refer to this tissue [20]. The NALT is the first contact site of airborne antigens and has been proposed to have the ability to induce and regulate antigen-specific T_H1 or T_H2 mediated responses and B cell immune responses [19]. The major antibody produced when the NALT is stimulated is IgA [21].

In rats and mice, NALT is found at the ventral portion of the lateral walls of the nasopharyngeal duct (Fig. 9.3). This area is found at the end of level II and extends through level III, extending to the middle region of the second molar [12]. Sections from this area will contain ethmoid turbinates and the large nasopharyngeal duct. Serial sections may be required to evaluate all areas of NALT, as the T cell and B cell areas are distinct but not in specific fixed locations in different areas of NALT [12].

The follicle-associated epithelium that overlies the NALT is composed of non-ciliated cuboidal epithelium with luminal microvilli (microfold or M cells) with a few ciliated cuboidal cells and a few mucous cells [22]. Regarding the composition of the NALT for the rat and the mouse, the areas of B cells and T cells are approximately equivalent, but B cells are generally abundant, with reported B cell to T cell ratios varying between 3.8:1 and 0.9:1 based on mouse strain [8, 21, 23, 24]. T-helper cells are in greater number than cytotoxic T cells for rats and mice; in BALB/c mice, the T-helper to cytotoxic T cell ratio has been reported as 3.7:1 [23, 25]. Macrophages and dendritic cells are also present, but in much smaller numbers compared to Peyer's patches in the intestine [8]. The basal lamina of the NALT can be normally infiltrated by lymphocytes and macrophages [12]. Germinal centers are not appreciated in NALT but have been reported with antigen administration [8, 24]. High endothelial venules are present (HEV), although NALT HEVs express peripheral

Fig. 9.3 Cross section of
the skull and nasal
turbinates, mouse.
Nasal-associated lymphoid
tissue is circled. R, oral
cavity; S, haired skin.
Hematoxylin and eosin, ×2

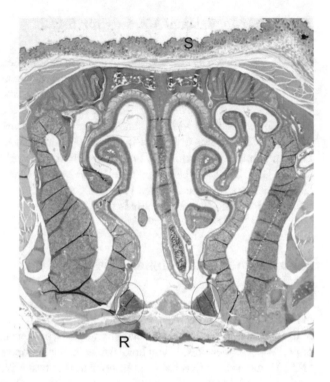

node addressin in contrast to the HEVs of Peyer's patches, that express mucosal
vascular addressin cell adhesion molecule 1 [9].

Hyperplastic Lesions

Airborne antigens often increase the size of NALT; however, formation of germinal
centers is infrequent [12]. Additionally, systemic administration of specific compounds that cause apoptosis and necrosis of lymphoid tissues can also cause apoptosis and necrosis in NALT [12]. Granulomatous inflammation and macrophage
aggregates have been observed in other regions of MALT, but not within the NALT
[16]. NALT drains the nasal mucosa and the olfactory nerves; therefore, tumor
emboli in the NALT may occur from primary tumors in these draining areas [16].

Immunohistochemical Markers for NALT

Many antibodies have been used in labeling lymphocytes and other cells within
NALT in various studies; however, the consistency in many of these antibodies has
not been examined, and not all work well in formalin-fixed tissues. Murine B cells

are reported to be consistently labeled with B220 (CD45R/B220), and this antibody has been used in NALT [17, 19, 26]. A consistent pan T cell marker in mice is CD3, although other studies have used CD5 in NALT to detect T cells, but this marker only detects a subset of T cells [19, 26]. Differentiation of T cells can be done with CD4 and CD8 markers; however, current antibodies for CD4 and CD8 do not work well in formalin-fixed tissues, and therefore, CD4 and CD8 cells must be identified by flow cytometry or immunohistochemistry of frozen section [19]. CD 138 labels plasma cells and Mac-2, Mac-3, and lysozyme are consistent markers for macrophages, although other studies have used Mac-1 (CD11b) with success on frozen sections for macrophages in NALT [17, 19, 25]. Antibodies for IgG, IgM, IgA, and IgE have been used with success in rat NALT to further differentiate B cells [16, 19].

References

1. Fry MM, McGavin MD. Bone marrow, blood cells, and lymphatic system. In: Zachary JF, McGavin MD, editors. Pathologic basis of veterinary disease. 5th ed. St. Louis, MO: Elsevier; 2012. p. 698–770.
2. Valli VE. Normal and benign reactive hematopoietic tissues. In: Veterinary comparative hematopathology. Oxford: Blackwell; 2007. p. 9–39.
3. Ward JM, et al. Thymus, spleen, and lymph nodes. In: Maronpot RR, Boorman GA, Gaul BW, editors. Pathology of mouse: reference and atlas. Vienna, IL: Cache River; 1999. p. 333–60.
4. Brown CC, Baker D, Barker IK. Alimentary system. In: Maxie MG, editor. Pathology of domestic animals. 5th ed. Edinburgh: Saunders; 2007. p. 1–296.
5. Sante-marie G, Peng FS. Morphological anomalies associated with immunodeficiencies in the lymph nodes of aging mice. Lab Invest. 1987;56:598–610.
6. Casola S, Rajewsky K. B cell recruitment and selection in mouse GALT germinal centers. Curr Top Microbiol Immunol. 2006;308:155–71.
7. Adachi S, Yoshida H, Kataoka H, Nishikawa S. Three distinctive steps in Peyer's patch formation of murine embryo. Int Immunol. 1997;9(4):507–14.
8. Cesta M. Normal structure, function, and histology of mucosa-associated lymphoid tissue. Toxicol Pathol. 2006;34:599–608.
9. Kiyono H, Fukuyama S. NALT- versus Peyer's-patch-mediated mucosal immunity. Nat Rev. 2004;4:699–710.
10. Bruder MC, et al. Intestinal T lymphocytes of different rat strains in immunotoxicity. Toxicol Pathol. 1999;27:171–9.
11. Sminia T, Kraal G. Nasal associated lymphoid tissue. In: Pearay OL, Mestecky J, Lamm ME, Strober W, Bienenstock J, McGhee JR, editors. Mucosal immunity. San Diego, CA: Academic; 1999. p. 357–64.
12. Elmore S. Enhanced histopathology of mucosa-associated lymphoid tissue. Toxicol Pathol. 2006;34:687–96.
13. Andoh A, et al. Serum antibody response and nasal lymphoid tissue (NALT) structure in the absence of IL-4 or IFN-γ. Cytokine. 2002;3:107–12.
14. Daniels CK, Perez P, Schmucker DL. Alterations in CD8+ cell distribution in gut-associated lymphoid tissues (GALT) of the aging Fischer 344 rat: a correlated immunohistochemical and flow cytometric analysis. Exp Gerontol. 1993;28:549–55.
15. Verschuere S, Bracke KR, Demoor T, et al. Cigarette smoking alters epithelial apoptosis and immune composition in murine GALT. Lab Invest. 2011;91(7):1056–67.
16. Kuper C. Histopathology of mucosa-associated lymphoid tissue. Toxicol Pathol. 2006; 34:609–15.

17. Ward J, et al. Immunohistochemical markers for the rodent immune system. Toxicol Pathol. 2006;34:616–30.
18. Greaves P. Histopathology of preclinical toxicity studies: interpretation and relevance in drug safety evaluation. 3rd ed. Oxford, UK: Elsevier; 2007.
19. Kunder S. A comprehensive antibody panel for immunohistochemical analysis of formalin-fixed, paraffin-embedded hematopoietic neoplasms of mice: analysis of mouse specific and human antibodies cross-reactive with murine tissue. Toxicol Pathol. 2007;35:366–75.
20. Casteleyn C. The tonsils revisited: review of the anatomical location and histological characteristics of the tonsils of domestic and laboratory animals. Clin Dev Immunol. 2011; 2011:472460.
21. Zuercher A, et al. Nasal-associated lymphoid tissue is a mucosal inductive site for virus-specific humoral and cellular immune responses. J Immunol. 2002;168:1796–803.
22. Harkema JR, Carey SA, Wagner JG. The nose revisited: a brief review of the comparative structure, function, and toxicology pathology of the nasal epithelium. Toxicol Pathol. 2006; 34:252–69.
23. Kuper CF. Lymphoid and non-lymphoid cells in nasal-associated lymphoid tissue (NALT) in the rat: an immuno- and enzyme-histochemical study. Cell Tissue Res. 1990;259:371–7.
24. Liang B, et al. Nasal-associated lymphoid tissue is a site of long-term virus-specific antibody production following respiratory virus infection of mice. J Virol. 2001;75:5416–20.
25. Heritage P. Comparison of murine nasal-associated lymphoid tissue and Peyer's patches. Am J Respir Crit Care Med. 1997;156:1256–62.
26. Stefanski SA, Elwell MR, Stromberg PC. Spleen, lymph nodes, and thymus. In: Boorman GA et al., editors. Pathology of Fischer rat. New York: Academic; 1990. p. 369–93.

Chapter 10
Bone Marker and Immunohistochemistry Changes in Toxic Environments

Ahmad Nazrun Shuid, Isa Naina Mohamed, Norliza Muhammad, Elvy Suhana Mohd Ramli, and Norazlina Mohamed

Introduction

Bone is a living tissue which is adversely affected by toxic environment. Bone cells are made up of osteoblasts and osteoclasts, which are responsible for bone formation and resorption, respectively. There is also osteocyte, which is actually old osteoblast that has lost its function. The activities of osteoblasts and osteoclasts play an important role in bone remodelling by maintaining the balance between bone resorption and formation. This balance may be disturbed when the bone cells are exposed to toxic environment. In some cases, the osteoblast function is adversely affected due to apoptosis or reduced activity in the presence of toxic environment, while in other cases, the activity or formation of osteoclast is increased. Both events would lead to bone loss and if severe enough, osteoporosis.

There are several conditions which may induce bone toxicity. Endotoxins released during bacterial infections may stimulate certain bone-resorbing cytokines or proteases. Oestrogen deficiency and hyperglycaemia are examples of endocrine disturbances which may lead to bone toxicity. Chronic use of steroids and nonsteroidal anti-inflammatory drugs or several other drugs such as tenofovir and isotretinoin may also cause bone toxicity. Drug addiction such as heroin or morphine and substance abuse such as alcohol and nicotine are associated bone loss. Even certain chemicals in diet or drinks such as caffeine in coffee or phosphoric acid in carbonated drinks are thought to be hazardous to bone health, especially if taken in excessive amount. Metals such as iron, lead and cadmium may be toxic to bone cells, while fluoride is a mineral known to cause bone toxicity.

A.N. Shuid (✉) • I.N. Mohamed • N. Muhammad • E.S.M. Ramli • N. Mohamed
Bone Metabolism Research Group, Department of Pharmacology,
Faculty of Medicine UKM, Level 17, Preclinical Building, JlnYaacob Latiff,
Bdr Tun Razak, 56000 Kuala Lumpur, Malaysia
e-mail: anazrun@yahoo.com; isanaina@yahoo.co.uk; norliza_ssp@yahoo.com;
elvysuhana@yahoo.com.my; azlina@medic.ukm.my

© Springer Science+Business Media New York 2016
S.A. Aziz, R. Mehta (eds.), *Technical Aspects of Toxicological Immunohistochemistry*, DOI 10.1007/978-1-4939-1516-3_10

When bones are exposed to toxic condition, they produced certain proteins or chemicals which can act as bone markers to indicate bone toxicity. These markers can be detected with immunohistochemistry technique or by using other laboratory techniques. Immunohistochemistry refers to the process of detecting proteins including markers in bone tissue section by applying the principle of antibodies binding specifically to antigens in bone tissues. It is commonly used in basic research to understand the distribution and localisation of biomarkers and differentially expressed proteins in different parts of a biological tissue. In this chapter, bone marker changes, with emphasis on detection by immunohistochemistry, were discussed when bone is exposed to toxic environment.

Endocrine Changes

Oestrogen Deficiency

Oestrogen deficiency has been known to cause bone loss and osteoporosis. There are several proposed mechanisms behind the bone loss associated with oestrogen deficiency. These include an increase free radical activity or oxidative stress induced by oestrogen deficiency. Oestrogen itself was found to act as an antioxidant as it was able to offer antioxidant protection for lipoproteins [1]. Oestrogen deficiency may expose the bone to free radical toxicity through the activation of nuclear factor kappa B (NF-κB) [2].

In a recent study, the effects of oestrogen deficiency on bone morphogenetic protein (BMP)-2, a protein involved in bone development, were determined via immunohistochemistry in ovariectomised rat, a postmenopausal osteoporosis model [3]. Sections of the calvarial bone of ovariectomised rats were labelled with human-specific anti-BMP-2 antibody, and the immunolocalised BMP-2 in the trabecular bone was quantified. It was shown that the immunolocalised BMP-2 area was decreased in the ovariectomised rats compared to sham-operated rats. This BMP-2 reduction may be another mechanism of bone loss caused by oestrogen deficiency conditions.

Oestrogen deficiency seems to provide a toxic environment which leads to bone loss, and oestrogen replacement was able to reverse this unfavourable condition. An immunohistochemistry study on rat alveolar bone demonstrated that administration of oestrogen resulted in positive immunolabelling of oestrogen receptor (ER)-β in the cytoplasms of osteoblasts, osteocytes and osteoclasts [4]. The results provide clear evidences that bone cells were influenced by the presence or absence of oestrogen via their action on ER-β.

Hyperglycaemia and Diabetes

Diabetes is a chronic disease that arises when the pancreas failed to produce enough insulin or when the body cannot effectively use the insulin produced by the pancreas. This resulted in raised blood glucose or hyperglycaemia which caused damage to several organ systems including the musculoskeletal system. Diabetes has been associated with bone loss, such as osteopenia and osteoporosis [5].

In most animal studies, streptozocin (STZ) was used as a diabetogenic agent to induce diabetes in rats. The STZ-induced diabetic rats were found to have high bone turnover, osteopenia and compromised bone strength [6–8]. The bone loss has been attributed to the prolonged hyperglycaemic condition in diabetes.

A study was carried out in STZ-induced diabetic rats to determine the effects of diabetes on several matrix-degrading enzymes (collagenase, gelatinase, elastase and beta-glucuronidase) in the gingiva of rats. Results showed that diabetes markedly increased the four matrix-degrading enzyme activities in the gingiva [9].

In another study, immunohistochemistry of the bone alkaline phosphatase (ALP) revealed that osteoblastic differentiation and maturation were reduced in streptozotocin-induced diabetic rats, but not to significant levels [10]. Serum ALP is normally increased when there is rapid bone loss, fracture and metabolic diseases such as diabetes mellitus [11].

Endotoxins

Lipopolysaccharides (LPS) are endotoxin found in the outer membrane of Gram-negative bacteria. LPS induces bone resorption in organ cultures by stimulating matrix metalloproteinase (MMP) from the osteoblast [12].

In a study, rats were injected with LPS (10 mg/ml) into the labial gingiva every other day for 3 weeks. It was found that the local LPS injection had increased the MMP activities in the gingiva [9]. In another study, rats were injected with LPS every other day for 6 days at the maxillary jaw. The maxilla was then extracted and processed for immunohistochemistry to detect the cytokines, including tumour necrosis factor (TNF), interleukin (IL)-1, IL-6, MMP-2 and MMP-9. The immuno-histochemistry results were found to be positive for TNF, IL-1 and IL-6 [13].This further illustrates that endotoxin may be responsible for bone loss by stimulating the cytokines and MMP levels.

Glucocorticoid

Human bone tissue expresses both reductase (cortisone to cortisol conversion) and dehydrogenase (cortisol to cortisone conversion) but it acts predominantly as a reductase in intact bone tissues. 11β-HSD1 reductase activity in osteoblasts generates active glucocorticoids in bone that are important for proliferation and differentiation of osteoblasts.

11β-Hydroxysteroid dehydrogenase type 1 (HSD1) enzyme acts as the local activator regenerating the natural ligand of growth receptor (GR) at bone level to produce the biological effect. 11β-HSD1 interconverts inactive cortisone into active

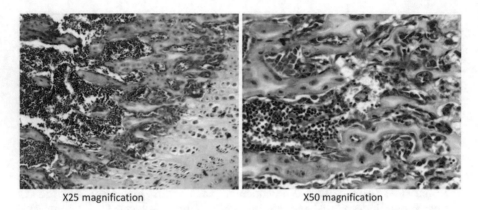

X25 magnification   X50 magnification

Fig. 10.1 The photomicrographs demonstrate immunohistochemical staining of 11beta-hydroxysteroid dehydrogenase type 1 (HSD1) activity in bone. HSD1 is an enzyme that interconverts inactive to active steroids. Samples were made up of sections of decalcified distal rat femur. HSD1 antibody was used to detect HSD1 enzyme. Immunopositive stain is shown by the *brown colour* staining

cortisol via pre-receptorial regulation. 11β-HSD1 is expressed in osteoblasts and plays a role in determining susceptibility of bone to glucocorticoid-induced osteoporosis. Bone tissue-specific response to glucocorticoids is strongly correlated to serum cortisone levels, not cortisol. Study has found that long-term glucocorticoids treatment reduced the dehydrogenase (active to inactive conversion) 11β-HSD1 activity in the bone which may increase the active glucocorticoids level that contributes to the development of osteoporosis. Immunohistochemistry study has also shown an increase in 11β-HSD1 expression in the bone after long-term glucocorticoid treatment [14]. In another immunohistochemistry study, it was shown that glucocorticoid was associated with increased Dmp1 and MMP-9 expressions and the local demineralisation around osteocytes [15] (Fig. 10.1).

Abused Substance

Substance of abuse has been known to affect human body both psychologically and physiologically. The drugs which are commonly abused can be categorised according to several groups: stimulants (cocaine, amphetamine, nicotine), opioids and depressants (alcohol, benzodiazepines) and miscellaneous (marijuana, ketamine). These substances of abuse have been associated with many negative effects on various organs in the body [16]. The use of these substances may also affect the bone which is often overlooked and ignored. In this section, the effects of three substance of abuse on bone markers were discussed.

Nicotine

Nicotine, the major addictive substance present in cigarette, has been implicated in many conditions such as cancers, cardiovascular diseases, lung diseases, pregnancy-related complications, negative effects on children and many other conditions [17, 18]. Several reports have been published with regard to the effects of nicotine on bone. It was noted that at low dose, nicotine exerted stimulatory effects on bone. However, high dose of nicotine was shown to negatively affect bone in both in vivo and in vitro studies [19].

In terms of bone mineral density, there were conflicting results. In a study, female rats were exposed to cigarette smoke for the duration of either 2, 3 or 4 months. It was observed that bone mineral density was reduced in rats exposed to 4 months of cigarette smoke [20]. In another study using female rats, it was shown that neither low dose (0.4 mg/kg/day) nor high dose of nicotine (6.0 mg/kg/day) given for 1 year in drinking water had any significant effects on bone mineral density [21]. The different route of administration and duration of exposure to nicotine may have affected bone turnover differently.

Despite the inconclusive findings on bone mineral density, several other studies were consistent in reporting the adverse effects of nicotine on bone markers. In an in vitro study on human mesenchymal stem cells, nicotine caused a dose-dependent reduction effect on cellular proliferation and expression of bone morphogenetic protein-2 (BMP-2), which is a marker for bone formation [22]. Similar findings on BMP-2 were observed in rabbits which were implanted with nicotine pellet subcutaneously for 7 weeks [23].

Bone formation marker (bone-specific alkaline phosphatase) was found to be reduced, while bone resorption marker (tartrate-resistant acid phosphatase 5b) was found to be increased in rats exposed to cigarette smoke for 3 or 4 months [20]. In another study, nicotine given intraperitoneally at the dose of 7 mg/kg to male rats for 2 months was shown to increase the levels of bone-resorbing cytokines (interleukin-1 and interleukin-6) and bone resorption marker (pyridinoline), while causing a decrease in osteocalcin (bone formation marker). It was also shown that cessation of nicotine for 2 months did not reverse the negative effects of nicotine on these bone biomarkers [17]. However, postmenopausal women who smoked at least ten cigarettes per day had higher bone mineral density after cessation of smoking for 1 year compared to women who continued smoking. Cessation of smoking was associated with increased alkaline phosphatase levels [24].

Besides the bone metabolism markers, nicotine was observed to negatively affect bone histomorphometry parameters [25]. Nicotine at the dose of 7 mg/kg given intraperitoneally for 2 months caused a decrease in bone volume, osteoblast surface, double-labelled surface, mineral appositional rate and bone formation rate. Nicotine was also found to decrease trabecular thickness and increased osteoclast surface and eroded surface [26].

Nicotine was also proven to delay bone healing. A study was carried out to evaluate the effects of nicotine on tendon-to-bone healing in rats. Proliferating cell nuclear antigen immunohistochemistry analysis found that the cellular prolif-eration in the nicotine group was decreased. It was also shown that nicotine caused prolonged inflammation as evidenced by the presence of inflammatory cells. These effects led to delayed healing process and thus reduction in biome-chanical strength of the bone [27]. The presence of tumour necrosis factor-α (TNF-α) was necessary to accelerate bone fracture healing and nicotine was found to inhibit the secretion of TNF-α, which may partly be responsible for causing the delay in healing [28].

Based on the above findings, it is undeniable that nicotine can cause detrimental effects on bone. However, nicotine may not be the sole culprit in delaying fracture healing. Other chemicals beside nicotine which are present in the cigarette smoke may also be responsible for these detrimental effects on bone [29].

Opioids

Opioids are a family of agents that are commonly used to relieve pain and are also often abused for its euphoric effects. Usage of opioid for long-term pain manage-ment is on the rise and concerns on its safety have been raised. The adverse effects of opioids on multiple organs may occur via various mechanisms [30]. Furthermore, long-term use of opioids may lead to endocrinopathy such as opioid-induced androgen deficiency [31] and its systemic effects including osteoporosis and osteopenia [32].

Prolonged heroin addicts were observed to have osteopenia at the femoral neck and distal forearm. These subjects also had increase bone turnover markers which were restored to normal levels after 1 year of methadone maintenance therapy [33]. Even though it seems that methadone maintenance therapy was able to reverse the effects of heroin addiction on bone turnover, long-term methadone therapy may also exert negative effects on bone. In a study population which has been on methadone maintenance therapy for a median of 11 years, it was found that the male subjects had lower bone mineral density as compared to control. However, this effect was not seen in female subjects [34]. The lack of effects on bone mineral density in women was also observed in another study involving perimenopausal women [35].

In another study which involved patients with heroin and buprenorphine addic-tion during the period of abstinence, it was found that ossalgia (bone pain) was one of the symptoms most prominently felt. It was also observed that the patients exhib-ited high clearance of calcium in the urine which was suggestive of high bone resorption [36].

Opioids have also been implicated in increasing fracture risk [37]. In a case-control study, it was found that risk of fracture was evident with acute use but not chronic use of opioid, which suggested the key role of acute central nervous system effects of opioids in fracture risk rather than the chronic opioid-induced hypogonadism [38]. As mentioned earlier, opioids have endocrinopathy effects, which may account for the adverse effects of opioids on bone. However, in one study, bone mineral density of the population studied showed lower bone mineral density regardless of their total testosterone blood levels [39].

In another perspective study, it was reported that the effects of opioids on bone may be attributed to the presence of opioid receptors on osteoblast cells. This was proven via reverse transcriptase polymerase chain reaction (RT-PCR) and immuno-histochemistry analysis which showed the presence of three types of opioid recep-tors. It was also found that osteocalcin synthesis was inhibited by high concentrations of the mu agonists morphine (D-Ala(2), N-MePhe(4), Gly(5)-ol)-enkephalin) but not with the delta agonist (D-Ala(2), D-leu(5))-enkephalin). These effects were abolished when osteoblast cells were incubated simultaneously with naloxone, an opioid antagonist [40]. The existence of these receptors and the varying affinity of individual opioids towards these receptors may account for the different effects of these opioids on the bone. Amongst the opioids commonly used, tramadol was observed to have the least effects on bone as compared to morphine and fentanyl as evidenced by the biochemical and histological parameters [41].

Alcohol

Besides the two substance of abuse discussed above, chronic alcohol abuse is estab-lished as an independent risk factor for osteoporosis and has many effects on bone mass, bone metabolism and bone strength [42].

There is a correlation between chronic alcohol abuse and low bone mass which was prominent in males [43]. Alcohol consumption was found to decrease fat mass and lean mass which are directly related to bone mineral density [44]. Alcohol was also able to reduce serum estradiol level via the increase in receptor activator of nuclear factor κB ligand (RANKL) expression, which eventually led to decrease in bone mineral density [45, 46]. Conflicting bone mineral density findings were observed in another study. Moderate alcohol consumption in postmenopausal women showed an increase in bone mineral density, which was associated with reduction in biochemical markers of bone turnover [47]. This suggests that the inhibitory effects of alcohol on the bone turnover attenuate the imbalance in bone turnover associated with menopause.

Alcohol was shown to exert direct effects on bone by reducing osteoblast activity and differentiation as well as osteoblastogenesis, which resulted in decrease bone formation [48]. In addition, alcohol increased bone resorption by increasing osteo-clastogenesis and osteoclast activity which was mediated by cytokines such as inter-leukin-6 (IL-6) and RANKL [49, 50].

Several studies have shown that alcohol intake in both human and rats resulted in reduced bone formation [51, 52]. Elevated blood alcohol concentrations for con-tinuous periods of time during repeated binge cycles in male rats were found to alter bone remodelling [43, 53, 54]. Consequently, there were reduced bone formation and increased resorption as a result of remodelling imbalance. Studies in young and adult rats have found that alcohol can affect osteogenesis, leading to altered cortical and cancellous architecture, decreased bone mineral density and significant reduction bone strength [55–57]. The increased blood alcohol concentrations due to binge alcohol administration in male rats were responsible for alcohol-induced osteoporosis mediated by bone formation inhibition and bone resorption stimulation [58].

Alcoholics were also associated with risk of fractures [59]. Exposure to alcohol delayed fracture healing by disrupting the canonical Wnt signalling which plays a major role in fracture repair [60]. Exposure to different type of alcohol may affect the outcome of the fracture risk. Women who preferred wine had lower hip fracture risk compared to the other alcohol drinkers [61].

Phosphoric Acid in Carbonated Drinks

Humans consume approximately 1500 mg of phosphorus daily from unprocessed foods, such as milk, cheese, beef, poultry, fish, etc. [62]. Consumption of phosphorus is further increase to more than double in the past few decades with increasing popularity of carbonated drinks and processed foods. Consumption of carbonated beverages continue to increase and is currently the preferred beverage for young adults worldwide [63]. Carbonated beverages contain phosphoric acid. The addition of phosphates in food processing has now become common practice, and this has also added to the intake of phosphate in humans [64]. The intake of phosphorus and phosphoric acid has been hypothesised to cause bone loss in humans. Although concrete association between high phosphorus intake and bone loss in humans is still being debated [65–67], various rat models for research have since been used to illicit this association.

Since the early 1970s, various studies have demonstrated that an increase in phosphorus intake would cause an increase in bone loss [63, 68–70]. This is especially true if the ratio of phosphorus to calcium intake was more than two times higher. Increasing the calcium intake may diminish the calcium loss. However, continuing increase of both phosphorus and calcium even with a ratio of 1:1 resulted in bone loss, indicating that the compensation of increased calcium intake to reduce bone loss due to phosphorus has its limitation [70, 71]. Draper et al. [71] demonstrated that osteoporosis can be induced in aging rats by high phosphorus diets. Calcium loss from skeleton, measured by calcium excretion, was increased by providing more than 0.3 % of phosphorus in the diet. Rats fed 0.6 %, 1.2 % or 1.8 % phosphorus diet had an average loss of 17.6 %, 40.2 % and 34.2 % of calcium, respectively, from bone when compared with rats fed with 0.3 % phosphorus. Garcia-Contreras et al. [63] further elicited that rats given cola soft drink had significantly lower femoral BMD when compared with rats given water. Histological femoral cortical width was also significantly lower amongst rats given cola soft drink.

There are a few postulated mechanisms regarding bone loss due to increase phosphorus intake in rats. Draper et al. [69, 71, 72], in a series of three papers, postulated that high phosphorus intake will result in secondary hyperparathyroidism, which leads to increase in bone resorption. This hypothesis was strengthened with the demonstration that bone loss does not occur in parathyroidectomised rats [68].

Later, it was found that phosphorus intake would inhibit renal 1α, 25-dihydroxyvitamin-D synthesis which blocks intestinal absorption and renal tubular reabsorption of calcium [73]. Increase phosphorus intake will therefore result in hypocalcaemia, secondary hyperparathyroidism and reduction in 1α, 25-dihydroxyvitamin-D level.

High phosphorus diet induces osteoporosis in rats. It can be safely assumed that with rats fed on high phosphorus diet, immunohistochemistry changes consistent with osteoporosis should be seen.

Caffeine

There were conflicting evidences on the association of caffeine intake with bone loss and increase risk of fracture in humans [74]. Several observational studies suggested a reduction in bone mineral density (BMD) with caffeine intake [75–77], while many others suggested no association [78–82]. It can be concluded that caffeine intake may have an effect on bone loss or calcium loss, but this effect can be easily offset by optimal intake of calcium according to recommended daily intake [74].

The effect of caffeine on bone structure yields similar conflicting results. Glajchen et al. [83] reported no effect on bone remodelling in adult rats, even at very high levels of caffeine consumption daily for 8 weeks. Sakamoto et al. [84] also concluded no effect of bone metabolism in adult rats given high coffee diets. However, a recent study by Liu et al. [85] demonstrated that caffeine enhances COX 2/PGE$_2$-regulated RANKL-mediated osteoclastogenesis in cell culture model. Three-week-old rats fed with caffeine for 20 weeks demonstrated significantly lower bone mineral density of lumbar vertebra, femur and tibia compared with the control group. Calcium contents were also lower in the tibia and femur bones of the rats [85].

It is possible that caffeine intake in rats may cause bone loss changes in rat bones. However, the evidence is currently inconclusive and more research in this area is needed.

Metal Toxicity

Iron

Iron overload is a condition where there is accumulation of iron in the body. The most important cause of iron overload is a disease called hemochromatosis. In this disease, the body tends to absorb high amounts of iron from the diet and the excess iron is stored in organs, especially the liver, heart and pancreas.

Another condition which can lead to iron overload is by repeated blood transfusion. Each blood transfusion increases the iron level in the body. As the red cells break down over time, the iron in the haemoglobin is released. However, the body has no natural way to remove the excess iron. Therefore, a protein called transferrin carries iron for storage in various organs including the bone marrow. As the bone marrow is in close vicinity to the bone and the bone cells are derived from the progenitor cells in bone marrow, the excess iron could cause bone toxicity. This may lead to bone loss, which is reflected by changes in certain bone markers.

Recent human studies have shown that iron overload may be toxic to bone. In a study by Kim et al. [86], the iron level was measured in relation to changes in bone mineral density of several bone sites. It was proven that iron stores could be an independent risk factor for accelerated bone loss, even in healthy populations. In another study, the relationships between iron levels and bone metabolism in patients with hip fragility fractures were examined [87]. There was a high prevalence of iron overload in postmenopausal women with fragility fracture. Iron overload may lower bone mineral density and cause bone loss by enhancing the activity of bone turnover in postmenopausal women. Another proposed mechanism is that iron generates free radicals, which are toxic to bone cells. The toxic effects of iron could be assessed by measuring the bone turnover markers such as serum osteocalcin and C-terminal telopeptide (CTX) or other markers such as the bone-resorbing cytokines.

The toxic effects of iron overload on bone have been investigated in rodent models. One of the widely accepted iron overload models is the ferric-nitrilotriacetate (FeNTA) model. In this model FeNTA, a compound made up of ferum chelated with nitrilotriacetate, is injected intraperitoneally into rats. Ferum or iron is a strong oxidant, while nitrilotriacetate stabilises the compound. It is the iron component of FeNTA which generates free radicals through Fenton reaction [88]. Studies have shown that free radicals were toxic to bone and may cause a reduction in bone mineral density [89]. Sontakke et al. [90] have reported that subjects with high bone loss have elevated malondialdehyde level, an index of lipid peroxidation, while the antioxidant enzymes were reduced. In another study, Maggio et al. [91] showed that elderly women suffering from bone loss have low levels of natural antioxidants. Therefore, free radicals may be responsible for the pathogenesis of osteoporosis.

Several studies using the FeNTA rat model have demonstrated the toxic effects of iron overload and free radicals on bone. Iron was found to be deposited in bone cells and there was an increase in the number of osteoclast, the cells responsible for bone resorption [92]. Other negative effects of FeNTA include deterioration of the structural, static and dynamic parameters of bone histomorphometry [93], impairment of bone mineralisation [94] and decreased bone calcium [95]. In terms of the bone markers, FeNTA was shown to increase the levels of bone-resorbing cytokines, interleukin-1 and interleukin-6 accompanied by elevated levels of deoxypyridinoline cross-link, a bone marker of osteoclast activity [96].

Isomura et al. [97] have investigated the relationship between free radicals induced by iron and bone metabolism in young and postmenopausal female rats. This was achieved by giving dietary iron overload to rats and measuring several bone metabolic, oxidative stress and antioxidant markers. The markers were measured using enzyme-linked immunosorbent assay (ELISA) technique, automated analyser or enzyme assay. Postmenopausal rats exhibited reduction in both the bone formation markers of serum alkaline phosphatase activity and osteocalcin level. However, there was no difference in the urinary excretion of deoxypyridinoline (DPD), a bone resorption marker. Iron-rich diet in postmenopausal rats has led to high excretion of DPD and 8-hydroxy-2'-deoxyguanosine, a marker of oxidative stress.

The iron-rich diet induced significant increases of serum osteopontin and tumour growth factor (TGF)-β1, resulting in augmented receptor activator of nuclear factor kappa B ligand (RANKL)-induced osteoclast formation [98–100].

TGF-β1 showed a negative correlation with serum glutathione peroxidase (GPx) activity, but a positive correlation with the serum iron level. The study has clearly shown that free radicals were toxic to bone and may result in osteoporosis, in which the concentrations of biochemical markers for oxidative stress and bone resorption were correlated to each other.

The opinion that bone toxicity of iron overload may be contributed by free radicals was supported by findings that antioxidants may reverse the toxicity. Vitamin E is an example of a lipid-soluble antioxidant which could suppress lipid peroxidation by breaking the peroxyl radical chain reaction. Vitamin E in the form of alpha-tocopherol was shown by Ebina et al. [93] to protect bone from FeNTA toxicity, while Yee and Ima Nirwana [95] had shown that palm oil-derived vitamin E could do the same. Palm vitamin E was shown to suppress interleukin (IL)-1 elevation and bone turnover marker changes induced by FeNTA [96].

Cadmium

Cadmium, a by-product of mining and smelting industry, is a highly toxic metal which exerts disruptive effects to the body at doses much lower than most other toxic elements. It is mainly used in the production of alloys and rechargeable batteries. In polluted areas, cadmium is released into rivers and remains in soil for several decades, taken up by plants and contaminates the food chains. It is also present in tobacco smoke. Acute and chronic cadmium intoxications can occur by inhalation and ingestion. Absorbed cadmium irreversibly accumulates in organs such as kidneys, lungs, the liver and bones [101, 102].

The "Itai-itai" (which literally translates to "it hurts-it hurts") disease was the world's first documented mass cadmium poisoning that took place in Toyama Prefecture in Japan in 1912. A mining company in the industrialised Toyama Prefecture released cadmium into the local river and poisoned everyone who drank the water or ate foods from the plants which were watered with the toxic metal. The first symptoms of the disease appeared as spinal and leg pain, later progressed to skeletal deformities, bone fractures even from minimal trauma, anaemia and kidney problems [103].

The toxic effects of cadmium on bone during the outbreak of Itai-itai disease led to many laboratory and epidemiological studies. Recent findings suggest that even relatively low chronic exposure to this metal may pose a risk for bone diseases. Increased prevalence of osteopenia and osteoporosis as well as fractures has been noted at low to moderate environmental exposure to cadmium [104, 105]. A study involving Chinese farmers who were exposed to cadmium from contaminated rice from over 20 years showed that these farmers had decreased bone mineral density [106]. Staessen et al. [107] reported that in subjects with high cadmium level in their urine, the risk for bone fracture was higher. Studies in laboratory animals also showed similar results. In rats treated with cadmium-infused drinking water, the BMD and skeletal mineral contents were lower than those of control rats. The cadmium-induced skeletal changes resulted in the weakening of bone biomechanical properties [108].

The detrimental effects of cadmium on bone occur through direct bone damage and indirectly due to renal dysfunction, which begins as renal tubular proteinuria, calciuria and phosphaturia. The disturbances in calcium and phosphate homeostasis that accompany cadmium nephropathy may lead to bone demineralisation and fractures [101]. In culture systems, cadmium acts directly on bone cells to retard bone formation and increase bone resorption. In a mouse osteoblast-like cell line, cadmium exposure led to decreased cellular alkaline phosphatase [109] and increased prostaglandin E2 secretion; the latter can stimulate formation and activation of osteoclasts, leading to uncoupling of bone remodelling [110]. As for bone-resorbing ability, cadmium has been shown to increase the number and activity of multinucleated osteoclast-like cells formed by the fusion of mononuclear precursors in primary rat osteoclast cultures [111]. Urinary markers for bone demineralisation were assessed in a study involving a group of school children exposed to cadmium in Pakistan. In these children, those with high urinary cadmium were noted to have high urinary calcium and deoxypyridinoline, suggesting that cadmium exposure is associated with bone resorption [112].

Lead

Lead is a poisonous metal which occurs both in organic and inorganic forms in the environment [113]. It is present in the air, water, soil, food and in many household products including brass plumbing fixtures, paints and cooking and eating utensils [114]. Routes of exposure mainly occur through the respiratory and gastrointestinal systems. Lead is conjugated in the liver and excreted through the kidneys, but most of the absorbed lead accumulates in various organs particularly bones, lungs, heart, brain and testes.

Toxicity to lead occurs where the levels in the blood circulation are high. Lead poisoning can take place acutely, but chronic poisoning as a result of continuous exposure is more common, affecting many bodily functions at the molecular, cellular and intercellular levels [115]. Lead intoxication induces oxidative stress by increasing lipid peroxidation and decreasing the antioxidant system. The endogenous antioxidant enzymes such as superoxide dismutase, glutathione peroxidase, glutathione S-transferase and catalase enzyme levels were noted to be lowered.

In addition, lead toxicity also resulted in increased bilirubin and the enzymatic activity of glutamic-oxaloacetic transaminase, glutamate pyruvate transaminase and alkaline phosphatase [116–118].

Since bone is the major reservoir for body lead, the latter has profound detrimental effects to bone. Elevated lead in the body produces a characteristic dense metaphyseal lines on X-ray. In children, it has been associated with decreased stature and chest circumference [119, 120], along with reduced somatic growth, longitudinal bone growth and bone strength [121]. Lead exposure may lead to reduced bone density [122], osteopenia [123] and impaired bone mineralisation [124].

In vivo and in vitro studies showed that lead affects the bone by direct action on bone cell functions and indirectly via the changes in the circulating levels of the hormones and cytokines which modulate bone cells, particularly 1, 25-dihydroxyvitamin D3. High levels of lead in blood were associated with low level of 1, 25-dihydroxyvitamin D3 [125]. Lead induces chondrogenesis by altering multiple signalling pathways which include TGF-β, BMP, AP-1 and NFκB [126]. Lead adversely affects bone formation and resorption by targeting cellular pathways involving osteoblasts and osteoclasts. It significantly reduced the activity of alkaline phosphatase [127], type I collagen and osteocalcin [128]. Lead impairs skeletal healing in a mouse model by delaying cartilage mineralisation and formation of bridging cartilage, as well as decreased expressions of collagen type II and X.

Fluoride

Fluoride is a naturally occurring mineral which can be found in ground waters as well as in fruits, vegetables, teas and other crops. In the United States, the addition of fluoride to community waters began in the 1940s after scientists noted that people living in areas with higher water fluoride levels had fewer cavities in their teeth. Water fluoridation had since been practiced by governing bodies in many parts of the world. Studies showed that the average rate of tooth decay in children was greatly reduced to by 50 % when the children drank fluoridated water [129]. Because of the positive effects of fluoride on teeth, this mineral serves as essential component in toothpastes, mouth rinses and in many other commercial products aimed to prevent tooth decay.

Once in the body, fluoride is absorbed into the blood, distributed through the blood circulation, and tends to be accumulated in areas high in calcium, such as the bones and teeth. Chronic exposure to fluoride leads to detrimental effects on these organs. Beltrán-Aguilar et al. [130] reported that the reduction in dental caries correlates with increased prevalence of dental fluorosis, characterised by defects in the forms of white streaks, spots and pitting of dental enamel. Besides, fluoride toxicity also leads to skeletal fluorosis, which is hallmarked by osteosclerosis, calcifications of ligament, as well as symptoms of osteoporosis, osteomalacia or osteopenia [131]. In addition, current evidence strongly suggests that even in the absence of skeletal fluorosis, chronic fluoride ingestion would also result in osteoarthritis [132, 133].

The exact mechanism by which fluoride causes defects in teeth and skeleton is unknown, but evidences from in vitro and in vivo studies point out the various actions of fluoride on bone cells and extracellular matrix. A study using a bone mineralising culture system which analysed the expression of bone matrix metal-loproteinases (MPP) based on immunohistochemistry staining showed that fluoride alters the expression of MPP within the mineralising bone [134]. An in vivo study in rats showed excessive fluoride can inhibit the secretion of MMP-20 and disturb the balance between MMP-20 and tissue inhibitors of metalloproteinases (TIMP)-2 [135]. As a consequence, the composition of the remodelling matrix and the ensu-ing mineralisation process would change.

The actions of fluoride on bone cells particularly osteoblasts and osteoclasts have been well documented. In the early 1970s, it was discovered that fluoride induced new bone formation, albeit the poorly mineralised bone [136]. Later studies exam-ining the bone markers showed that fluoride stimulates bone formation by increas-ing alkaline phosphatase activity [137], bone sialoprotein and BMP-2 expression [138]. Other bone formation markers including osteocalcin, Runx2, and collage type I gene expression were also elevated [139]. Osteoclastogenesis was induced in C3H inbred mouse strain when fluoride was administered in drinking water for 3 weeks, as seen by elevated levels of PTH, RANKL and TRAP5b with a concordant decrease in serum OPG [140].

In a study whereby young rabbits were given high-dose fluoride in their drinking water for 6 months, the bone turnover rate was noted to increase, with marked eleva-tions of bone markers like bone alkaline phosphatase, serum TRAP and IGF-1. The study also showed that fluoride increased the vertebral volume by 35 % and tibial ash weight by 10 % [141].

The anabolic action of fluoride is confirmed by the increases in BMD, trabecular bone volume and trabecular thickness [142, 143]. However, the positive effects on bone formation were not reflected in improved bone strength. The study in rabbits by Turner et al. [141] noted that fluoride decreased bone strength by 19 % in the L5 vertebra and 25 % in the femoral neck. In a clinical trial involving postmenopausal women, it was shown that the fluoride-induced increase in BMD was associated with increased fracture risk [142]. The increases in bone volume and trabecular thickness caused by fluoride were not accompanied by a concomitant increase in trabecular connectivity. It is the lack of trabecular connectivity that results in poor bone quality and reduced bone strength [143].

Conclusion

Bone just like other live tissues can be adversely affected by toxic environments. Bone damage can be assessed by measuring certain markers that it releases using laboratory techniques such as immunohistochemistry. Therefore, bone toxicity can be detected or monitored for the purpose of prevention or treatment.

References

1. Badeau M, Adlercreutz H, Kaihovaara P, Tikkanen MJ. Estrogen A-ring structure and antioxidative effect on lipoproteins. J Steroid Biochem. 2005;96:271–8.
2. Iotsova V, Caamano J, Loy J, Yang Y, Lewin A, Bravo R. Osteopetrosis in mice lacking NF-kB1 and NF-kB2. Nat Med. 1997;3:1285–9.
3. Chen CH, Kang L, Lin RW, Fu YC, Li YS, Chang JK, Chen HT, Chen CH, Lin SY, Wang GJ, Ho ML. (-)-Epigallocatechin-3-gallate improves bone microarchitecture in ovariectomized rats. Menopause. 2013;20(6):687–94.
4. Cruzoé-Souza M, Sasso-Cerri E, Cerri PS. Immunohistochemical detection of estrogen receptorβ in alveolar bone cells of estradiol-treated female rats: possible direct action of estrogen on osteoclast life span. J Anat. 2009;215:673–81.
5. Räkel A, Sheehy O, Rahme E, LeLorier J. Osteoporosis among patients with type 1 and type 2 diabetes. Diabetes Metab. 2008;34:193–205.
6. Silva MJ, Brodt MD, Lynch MA, et al. Type 1 diabetes in young rats leads to progressive trabecular bone loss, cessation of cortical bone growth, and diminished whole bone strength and fatigue life. J Bone Miner Res. 2009;24:1618–27.
7. Horcajada-Molteni MN, Chanteranne B, Lebecque P, et al. Amylin and bone metabolism in streptozotocin induced diabetic rats. J Bone Miner Res. 2001;16:958–65.
8. Erdal N, Gürgül S, Kavak S, Yildiz A, Emre M. Deterioration of bone quality by streptozotocin (STZ)-induced type 2 diabetes mellitus in rats. Biol Trace Elem Res. 2011;140:342–53.
9. Chang KM, Ryan ME, Golub LM, Ramamurthy NS, McNamara TF. Local and systemic factors in periodontal diseases increase matrix-degrading enzyme activities in rat gingiva: effect of micocycline therapy. Res Commun Mol Pathol Pharmacol. 1996;91(3):303–18.
10. Choi CW, Lee HR, Lim HK. Effect of Rubus coreanus extracts on diabetic osteoporosis by simultaneous regulation of osteoblasts and osteoclasts. Menopause. 2012;19(9):1043–51.
11. Ross PD, Kress BC, Parson RE. Serum bone alkaline phosphatase and calcaneus bone density predict fractures: a prospective study. Osteoporos Int. 2000;11:76–82.
12. Sismey-Durrant HJ, Atkinson SJ, Hopps RM, Heath JK. The effect of lipopolysaccharide from bacteroides gingivalis and muramyl dipeptide on osteoblast collagenase release. Calcif Tissue Int. 1989;44(5):361–3.
13. Ramamurthy NS, Rifkin BR, Greenwald RA, Xu JW, Liu Y, Turner G, Golub LM, Vernillo AT. Inhibition of matrix metalloproteinase-mediated periodontal bone loss in rats: a comparison of 6 chemically modified tetracyclines. J Periodontol. 2002;73(7):726–34.
14. Elvy Suhana MR, Ima Nirwana S, Faizah O, Fairus A, Ahmad Nazrun S, Norazlina M, Norliza M, Farihah HS. The effects of Piper sarmentosum water extract on the expression and activity of 11β-hydroxysteroid dehydrogenase type 1 in the bones with excessive glucocorticoids. Iran J Biomed Sci. 2012;37(1):39–46.
15. Lane NE, Yao W, Balooch M, Nalla RK, Balooch G, Hebelitz S, et al. Glucocorticoid-treated mice have localized changes in trabecular bone mineral properties and osteocyte lacunar size that are not observed in placebo treated or estrogen-deficient mice. J Bone Miner Res. 2006;21:466–76.
16. Milroy CM, Parai JL. The histopathology of drugs of abuse. Histopathology. 2011;59(4):579–93.
17. Norazlina M, Hermizi H, Faizah O, Nazrun AS, Norliza M, Ima-Nirwana S. Vitamin E reversed nicotine-induced toxic effects on bone biochemical markers in male rats. Arch Med Sci. 2010;6(4):505–12.
18. Grief SN. Nicotine dependence: health consequences, smoking cessation therapies, and pharmacotherapy. Prim Care. 2011;38(1):23–39.
19. Kallala R, Barrow J, Graham SM, Kanakaris N, Giannoudis PV. The in vitro and in vivo effects of nicotine on bone, bone cells and fracture repair. Expert Opin Drug Saf. 2013;12(2):209–33.

20. Gao SG, Li KH, Xu M, Jiang W, Shen H, Luo W, Xu WS, Tian J, Lei GH. Bone turnover in passive smoking female rat: relationships to change in bone mineral density. BMC Musculoskelet Disord. 2011;12:131.
21. Turan V, Mizrak S, Yurekli B, Yilmaz C, Ercan G. The effect of long-term nicotine exposure on bone mineral density and oxidative stress in female Swiss Albino rats. Arch Gynecol Obstet. 2013;287(2):281–7.
22. Kim DH, Liu J, Bhat S, Benedict G, Lecka-Czernik B, Peterson SJ, Ebraheim NA, Heck BE. Peroxisome proliferator-activated receptor delta agonist attenuates nicotine suppression effect on human mesenchymal stem cell-derived osteogenesis and involves increased expression of heme oxygenase-1. J Bone Miner Metab. 2013;31(1):44–52.
23. Ma L, Zheng LW, Sham MH, Cheung LK. Uncoupled angiogenesis and osteogenesis in nicotine-compromised bone healing. J Bone Miner Res. 2010;25(6):1305–13.
24. Oncken C, Prestwood K, Kleppinger A, Wang Y, Cooney J, Raisz L. Impact of smoking cessation on bone mineral density in postmenopausal women. J Womens Health (Larchmt). 2006;15(10):1141–50.
25. Hapidin H, Othman F, Soelaiman IN, Shuid AN, Luke DA, Mohamed N. Negative effects of nicotine on bone-resorbing cytokines and bone histomorphometric parameters in male rats. J Bone Miner Metab. 2007;25(2):93–8.
26. Hapidin H, Othman F, Soelaiman IN, Shuid AN, Mohamed N. Effects of nicotine administration and nicotine cessation on bone histomorphometry and bone biomarkers in Sprague-Dawley male rats. Calcif Tissue Int. 2011;88(1):41–7.
27. Galatz LM, Silva MJ, Rothermich SY, Zaegel MA, Havlioglu N, Thomopoulos S. Nicotine delays tendon-to-bone healing in a rat shoulder model. J Bone Joint Surg Am. 2006;88(9): 2027–34.
28. Chen Y, Guo Q, Pan X, Qin L, Zhang P. Smoking and impaired bone healing: will activation of cholinergic anti-inflammatory pathway be the bridge? Int Orthop. 2011;35(9):1267–70.
29. Al-Hadithy N, Sewell MD, Bhavikatti M, Gikas PD. The effect of smoking on fracture healing and on various orthopaedic procedures. Acta Orthop Belg. 2012;78(3):285–90.
30. Baldini A, Von Korff M, Lin EH. A review of potential adverse effects of long-term opioid therapy: a practitioner's guide. Prim Care Companion CNS Disord. 2012;14(3). pii: PCC.11m01326.
31. Smith HS, Elliott JA. Opioid-induced androgen deficiency (OPIAD). Pain Physician. 2012;15(3 Suppl):ES145–56.
32. Brennan MJ. The effect of opioid therapy on endocrine function. Am J Med. 2013;126(3 Suppl 1):S12–8.
33. Wilczek H, Stěpán J. Bone metabolism in individuals dependent on heroin and after methadone administration. Cas Lek Cesk. 2003;142(10):606–8.
34. Grey A, Rix-Trott K, Horne A, Gamble G, Bolland M, Reid IR. Decreased bone density in men on methadone maintenance therapy. Addiction. 2011;106(2):349–54.
35. Vestergaard P, Hermann P, Jensen JE, Eiken P, Mosekilde L. Effects of paracetamol, non-steroidal anti-inflammatory drugs, acetylsalicylic acid, and opioids on bone mineral density and risk of fracture: results of the Danish Osteoporosis Prevention Study (DOPS). Osteoporos Int. 2012;23(4):1255–65.
36. Sikharulidze ZD, Kopaliani MG, Kilasoniia LO. Comparative evaluation of clinical symptoms and status of bone metabolism in patients with heroin and buprenorphine addiction in the period of withdrawal. Georgian Med News. 2006;(134):72–6.
37. Mattia C, Di Bussolo E, Coluzzi F. Non-analgesic effects of opioids: the interaction of opioids with bone and joints. Curr Pharm Des. 2012;18(37):6005–9.
38. Li L, Setoguchi S, Cabral H, Jick S. Opioid use for noncancer pain and risk of fracture in adults: a nested case-control study using the general practice research database. Am J Epidemiol. 2013;178(4):559–69.
39. Fortin JD, Bailey GM, Vilensky JA. Does opioid use for pain management warrant routine bone mass density screening in men? Pain Physician. 2008;11(4):539–41.

40. Pérez-Castrillón JL, Olmos JM, Gómez JJ, Barrallo A, Riancho JA, Perera L, Valero C, Amado JA, González-Macías J. Expression of opioid receptors in osteoblast-like MG-63 cells, and effects of different opioid agonists on alkaline phosphatase and osteocalcin secretion by these cells. Neuroendocrinology. 2000;72(3):187–94.
41. Boshra V. Evaluation of osteoporosis risk associated with chronic use of morphine, fentanyl and tramadol in adult female rat. Curr Drug Saf. 2011;6(3):159–63.
42. Mikosch P. Alcohol and bone. Wien Med Wochenschr. 2014;164(1–2):15–24.
43. Turner RT. Skeletal response to alcohol. Alcohol Clin Exp Res. 2000;24:1693–701.
44. Maddalozzo GF, Turner RT, Edwards CH, Howe KS, Widrick JJ, Rosen CJ, Iwaniec UT. Alcohol alters whole body composition, inhibits bone formation, and increases bone marrow adiposity in rats. Osteoporos Int. 2009;20:1529–38.
45. Ronis MJ, Wands JR, Badger TM, de la Monte SM, Lang CH, Calissendorff J. Alcohol-induced disruption of endocrine signaling. Alcohol Clin Exp Res. 2007;31:1269–85.
46. Chen JR, Lazarenko OP, Haley RL, Blackburn ML, Badger TM, Ronis MJ. Ethanol impairs estrogen receptor signaling resulting in accelerated activation of senescence pathways, whereas estradiol attenuates the effects of ethanol in osteoblasts. J Bone Miner Res. 2009;24:221–30.
47. Marrone JA, Maddalozzo GF, Branscum AJ, Hardin K, Cialdella-Kam L, Philbrick KA, Breggia AC, Rosen CJ, Turner RT, Iwaniec UT. Moderate alcohol intake lowers biochemical markers of bone turnover in postmenopausal women. Menopause. 2012;19(9):974–9.
48. Chavassieux P, Serre CM, Vergnaud P, Delmas PD, Meunier PJ. In vitro evaluation of dose-effects of ethanol on human osteoblastic cells. Bone Miner. 1993;22:95–103.
49. Cheung RC, Gray C, Boyde A, Jones SJ. Effects of ethanol on bone cells in vitro resulting in increased resorption. Bone. 1995;16:143–7.
50. Dai J, Lin D, Zhang J, Habib P, Smith P, Murtha J, Fu Z, Yao Z, Qi Y, Keller ET. Chronic alcohol ingestion induces osteoclastogenesis and bone loss through IL-6 in mice. J Clin Invest. 2000;106:887–95.
51. Sampson HW, Chaffin C, Lange JL, DeFee B. Alcohol consumption by young actively growing rats: a histomorphometric study of cancellous bone. Alcohol Clin Exp Res. 1997;21:332–9.
52. Nyquist F, Ljunghall S, Berglund M, Obrant K. Biochemical markers of bone metabolism after short and long time ethanol withdrawal in alcoholics. Bone. 1996;19:51–4.
53. Chakkalakal DA. Alcohol-induced bone loss and deficient bone repair. Alcohol Clin Exp Res. 2005;29:2077–90.
54. Wezeman FH, Emanuele MA, Emanuele N, Moskal 2nd SF, Woods M, Suri M, Steiner J, LaPaglia N. Chronic alcohol consumption during male rat adolescence impairs skeletal development through effects on osteoblast gene expression, bone mineral density, and bone strength. Alcohol Clin Exp Res. 1999;23:1534–42.
55. Peng TC, Kusy RP, Hirsch PF, Hagman JR. Ethanol induced changes in morphology and strength of femurs of rats. Alcohol Clin Exp Res. 1988;12:1655–9.
56. Hogan HA, Groves JA, Sampson HW. Long-term alcohol consumption in the rat affects femur cross-sectional geometry and bone tissue material properties. Alcohol Clin Exp Res. 1999;23:1825–33.
57. Wezeman FH, Juknelis D, Frost N, Callaci HJ. Spine bone mineral density and vertebral body height is altered by alcohol consumption in growing male and female rats. Spine. 2003;31:87–92.
58. Wezeman FH, Emanuele M, Moskal SF, Steiner J, Lapaglia N. Alendronate administration and skeletal response during chronic alcohol intake in the adolescent male rat. J Bone Miner Res. 2000;15:2033–41.
59. Kelly KN, Kelly C. Pattern and cause of fractures in patients who abuse alcohol: what should we do about it? Postgrad Med J. 2013;89(1056):578–83.
60. Lauing KL, Roper PM, Nauer RK, Callaci JJ. Acute alcohol exposure impairs fracture healing and deregulates β-catenin signaling in the fracture callus. Alcohol Clin Exp Res. 2012;36(12):2095–103.

61. Kubo JT, Stefanick ML, Robbins J, Wactawski-Wende J, Cullen MR, Freiberg M, Desai M. Preference for wine is associated with lower hip fracture incidence in post-menopausal women. BMC Womens Health. 2013;13:36.
62. Subar AF, Krebs-Smith SM, Cook A, Kahle LL. Dietary sources of nutrients among US adults, 1989 to 1991. J Am Diet Assoc. 1998;98:537–47.
63. Garcia-Conteras F, Paniagua R, Avila-Diaz M, Cabrera-Munoz L, Martinez-Muniz I, Foyo-Niembro E, Amato D. Cola beverages consumption induces bone mineralization reduction in ovariectomized rats. Arch Med Res. 2000;31:360–5.
64. Calvo MS. Dietary considerations to prevent loss of bone and renal function. Nutrition. 2000;16(7/8):564–6.
65. Fitzpatrick L, Heaney RP. Got soda? J Bone Miner Res. 2003;18(9):1570–2.
66. Heaney RP, Rafferty K. Carbonated beverages and urinary calcium excretion. Am J Clin Nutr. 2001;74:343–7.
67. Petridou E, Karpathios T, Dessypris N, Simou E, Trichopoulus D. The role of dairy products and non-alcoholic beverages in bone fractures among school age children. Scand J Soc Med. 1997;25:119.
68. Kim SH, Morton D, Barrett-Connor EL. Carbonated beverage consumption and bone density among older women: the Rancho Bernado Study. Am J Public Health. 1997;87:276.
69. Anderson GH, Draper HH. Effect of dietary phosphorus on calcium metabolism in intact and parathyroidectomized adult rats. J Nutr. 1972;102:1123–32.
70. Shah BG, Krishnarao GVG, Draper HH. The relationship of Ca and P Nutrition during adult life and osteoporosis in aged mice. J Nutr. 1967;92:30.
71. Draper HH, Sie TL, Bergan JG. Osteoporosis in aging rats induced by high phosphorus diets. J Nutr. 1972;102:1133–42.
72. Krishnarao GVG, Draper HH. Influence of dietary phosphate on bone resorption in senescent mice. J Nutr. 1972;102:1143–6.
73. Amato D, Maravilla A, Montoya C, Gaja O, Revilla C, Guerra R, Paniagua R. Acute effects of soft drink intake on calcium and phosphate metabolism in immature and adult rats. Rev Invest Clin. 1998;50:185.
74. Heaney RP. Effects of caffeine on bone and the calcium economy. Food Clin Toxicol. 2002;40:1263–70.
75. Bauer DC, Browner WS, Cauley JA, Orwoll ES, Scott JC, Black DM, Tao JL, Cummings SR. Factors associated with appendicular bone mass in older women. Ann Intern Med. 1993;118:657–65.
76. Krahe C, Friedman R, Gross JL. Risk factors for decreased bone density in premenopausal women. Braz J Med Biol Res. 1997;30:1061–6.
77. Rubin LA, Hawker GA, Peltekova VD, Fielding LJ, Ridout R, Cole DE. Determinants of peak bone mass: clinical and genetic analyses in a young female Canadian cohort. J Bone Miner Res. 1999;14:633–43.
78. Hannan MT, Felson DT, Dawson-Hughes B, Tucker KL, Cupples LA, Wilson PWF, Kiel DP. Risk factors for longitudinal bone loss in elderly men and women: the Framingham Osteoporosis study. J Bone Miner Res. 2000;15:710–20.
79. Picard D, Ste-Marie LG, Coutu D, Carrier L, Chartrand R, Lepage R, Fugere P, D'Amour P. Premenopausal bone mineral content relates to height, weight and calcium intake during early adulthood. Bone Miner. 1988;4:299–309.
80. Packard PT, Recker RR. Caffeine does not affect the rate of gain in spine bone in young women. Osteoporos Int. 1996;6:149–52.
81. Maini M, Brignoli E, Felicetti G, Bozzi M. Correlation between risk factors and bone mass in pre and postmenopause. Epidemiologic study on osteoporosis. Part 1. Minerva Med. 1996;87:385–99.
82. Grainge MJ, Coupland CA, Cliffe SJ, Chilvers CE, Hosking DJ. Cigarette smoking, alcohol and caffeine consumption, and bone mineral density in postmenopausal women. Osteoporos Int. 1998;8:355–63.

83. Glajchen N, Ismail F, Epstein S, Jowell PS, Fallon M. The effect of chronic caffeine administration on serum markers of bone mineral metabolism and bone histomorphometry in the rat. Calcif Tissue Int. 1988;43:277–80.
84. Sakamoto W, Nishihira J, Fujie K, Lizuka T, Handa H, Ozaki M, Yukawa S. Effect of coffee consumption on bone metabolism. Bone. 2001;28:332–6.
85. Liu SH, Chen C, Yang RS, Yen YP, Yang YT, Tsai C. Caffeine enhances osteoclast differentiation from bone marrow hematopoietic cells and reduces bone mineral density in growing rats. J Orthop Res. 2011;29:954–60.
86. Kim BJ, Ahn SH, Bae SJ, Kim EH, Lee SH, Kim HK, Choe JW, Koh JM, Kim GS. Iron overload accelerates bone loss in healthy postmenopausal women and middle-aged men: a 3-year retrospective longitudinal study. J Bone Miner Res. 2012;27(11):2279–90.
87. Zhang LL, Jiang XF, Ai HZ, Jin ZD, Xu JX, Wang B, Xu W, Xie ZG, Zhou HB, Dong QR, Xu YJ. Relationship of iron overload to bone mass density and bone turnover in postmenopausal women with fragility fractures of the hip. Zhonghua Wai Ke Za Zhi. 2013;51(6): 518–21.
88. Awai M, Naraski M, Yamanoi Y, Seno S. Induction of diabetes in animals by parenteral administration of ferric nitrilotriacetate. A model of experimental hemochromatosis. Am J Pathol. 1976;95:663–72.
89. Basu S, Michaelsson K, Olofsson H, Johansson S, Melhus H. Association between oxidative stress and bone mineral density. Biochem Biophy Res Comm. 2001;288(1):275–9.
90. Sontakke AN, Tare RS. A duality in the roles of reactive oxygen species with respect to bone metabolism. Clin Chim Acta. 2002;318(1–2):145–8.
91. Maggio D, Barabani M, Pierandrei M, Polidori MC, Catani M, Mecocci P, Senin U, Pacifici R, Cherubini A. Marked decrease in plasma antioxidants in aged osteoporotic women: results of a cross-sectional study. J Clin Endocrinol Metab. 2003;88(4):1523–7.
92. Yukihiro S, Okada S, Takeuchi K, Inoue H. Experimental osteodystrophy of chronic renal failure induced by aluminium and ferric nitrilotriacetate in Wistar rats. Pathol Int. 1995;45:19–25.
93. Ebina Y, Okada S, Hamazaki S, Toda Y, Midorikawa O. Impairment of bone formation with aluminum and ferric nitrilotriacetate complexes. Calcif Tissue Int. 1991;48(1):28–36.
94. Takeuchi K, Okada S, Yukihiro S, Inoue H. The inhibitory effects of aluminium and iron on bone formation, in vivo and in vitro study. Pathophysiology. 1997;4:97–104.
95. Yee JK, Ima Nirwana S. Palm Vit E protects bone against Ferric-Nitrilotriacetate-induced impairment of calcification. Asia Pac J Pharmacol. 1998;13:35–41.
96. Ahmad NS, Khalid BAK, Luke DA, Ima-Nirwana S. Tocotrienol offers better protection than tocopherol from free radical-induced damage of rat bone. Clin Exp Pharm Physiol. 2005;32:761–70.
97. Isomura H, Fujie K, Shibata K, Inoue N, Iizuka T, Takebe G, Takahashi K, Nishihira J, Izumi H, Sakamoto W. Bone metabolism and oxidative stress in postmenopausal rats with iron overload. Toxicology. 2004;197:93–100.
98. Fox SW, Haque SJ, Lovibond AC, Chambers TJ. The possible role of TGF-beta-induced suppressors of cytokine signaling expression in osteoclast/macrophage lineage commitment in vitro. J Immunol. 2003;170:3679–87.
99. Massey HM, Scopes J, Horton MA, Flanagan AM. Transforming growth factor-beta1 (TGF-beta) stimulates the osteoclast-forming potential of peripheral blood hematopoietic precursors in a lymphocyte-rich microenvironment. Bone. 2001;28:577–82.
100. Yamamoto N, Sakai F, Kon S, Morimoto J, Kimura C, Yamazaki H, Okazaki I, Seki N, Fujii T, Ueda T. Essential role of the cryptic epitope SLAYGLR within osteopontin in a murine model of rheumatoid arthritis. J Clin Invest. 2003;112:181–8.
101. Bernard A. Cadmium & its adverse effects on human health. Indian J Med Res. 2008;128:557–64.
102. Satarug S, Baker JR, Urbenjapol S, Haswell-Elkins M, Reilly PE, Williams DJ, Moore MR. A global perspective on cadmium pollution and toxicity in non-occupationally exposed population. Toxicol Lett. 2003;137:65–83.

103. Hagino N, Yoshioka Y. A study of the etiology of Itai-Itai disease. J Jpn Orthop Assoc. 1961;35:812–5.
104. Wang H, Zhu G, Shi Y, Wenig S, Jin T, Kong Q, Nordberg GF. Influence of environmental cadmium exposure on forearm bone density. J Bone Miner Res. 2003;18:553–60.
105. Kazantzis G. Cadmium, osteoporosis and calcium metabolism. Biometals. 2004;17:493–8.
106. Nordberg G, Jin T, Bernard A, Fierens S, Buchet JP, Ye T. Low bone density and renal dysfunction following environmental cadmium exposure in China. Ambio. 2002;31:478–81.
107. Staessen JA, Roels HA, Emelianov D, Kuznetsova T, Thijs L, Vangronsveld J, Fagard R. Environmental exposure to cadmium, forearm bone density, and risk of fractures: prospective population study. Public Health and Environmental Exposure to Cadmium (PheeCad) Study Group. Lancet. 1999;353(9159):1140–4.
108. Brzóska MM, Majewska K, Moniuszko-Jakoniuk J. Bone mineral density, chemical composition and biomechanical properties of the tibia of female rats exposed to cadmium since weaning up to skeletal maturity. Food Chem Toxicol. 2005;43(10):1507–19.
109. Miyahara T, Yamada H, Takeuchi M, Kozuka H, Kato T, Sudo H. Inhibitory effects of cadmium on in vitro calcification of a clonal osteogenic cell, MC3T3-E1. Toxicol Appl Pharmacol. 1988;96(1):52–9.
110. Bhattacharyya MH. Cadmium osteotoxicity in experimental animals: mechanisms and relationship to human exposures. Toxicol Appl Pharmacol. 2009;238(3):258–65.
111. Wilson AK, Cerny EA, Smith BD, Wagh A, Bhattacharyya MH. Effects of cadmium on osteoclast formation and activity in vitro. Toxicol Appl Pharmacol. 1996;140(2):451–60.
112. Sughis M, Penders J, Haufroid V, Nemery B, Nawrot TS. Bone resorption and environmental exposure to cadmium in children: a cross-sectional study. Environ Health. 2011;10:104.
113. Shalan MG, Mostafa M, Hassouna MM, Nabi SE, Rafie A. Amelioration of lead toxicity on rat liver with vitamin C and silymarin supplements. Toxicology. 2005;206:1–15.
114. Garaza A, Vega R, Soto E. Cellular mechanisms of lead neurotoxicity. Med Sci Monit. 2006;12(3):57–65.
115. Sidhu P, Nehru B. Lead intoxication: histological and oxidative damage in rat cerebrum and cerebellum. J Trace Elem Exp Med. 2004;17(1):45–53.
116. Elmenofi GAM. Bee honey dose-dependently ameliorates lead acetate-mediated hepatorenal toxicity in rats. Life Sci J. 2012;9(4):780–8.
117. Haleagrahara N, Chakravarthi S, Bangra Kulur A, Radhakrishnan A. Effects of chronic lead acetate exposure on bone marrow lipid peroxidation and antioxidant enzyme activities in rats. Afr J Pharm Pharmacol. 2011;5(7):923–9.
118. Kansal L, Sharma V, Sharma A, Lodi S, Sharma SH. Protective role of Coriandrum sativum (coriander) extracts against lead nitrate induced oxidative stress and tissue damage in the liver and kidney in male mice. Int J Appl Biol Pharm Technol. 2011;2(3):65–83.
119. Schwartz J, Angle C, Pitcher H. Relationship between childhood blood lead levels and stature. Pediatrics. 1986;77:281–8.
120. Shukla R, Bornschein RL, Dietrich KN, Buncher CR, Berger OG, Hammond PB, et al. Effects of fetal and infant lead exposure on growth in stature. Pediatrics. 1989;84:604–12.
121. Ronis MJ, Aronson J, Gao GG, Hogue W, Skinner RA, Badger TM, Lumpkin Jr CK. Skeletal effects of developmental lead exposure in rats. Toxicol Sci. 2001;62(2):321–9.
122. Escribano A, Revilla M, Hernandez ER, Seco C, Gonzalez-Riola J, Villa LF, et al. Effect of lead on bone development and bone mass: a morphometric, densitometric, and histomorphometric study in growing rats. Calcif Tissue Int. 1997;60:200–3.
123. Silbergeld EK, Sauk J, Somerman M, Todd A, McNeill F, Fowler B, et al. Lead in bone: storage site, exposure source, and target organ. Neurotoxicology. 1993;14:225–36.
124. Gangoso L, Alvarez-Lloret P, Rodriguez-Navarro AA, Mateo R. Long-term effects of lead poisoning on bone mineralization in vultures exposed to ammunition sources. Environ Pollut. 2009;157(2):569–74.
125. Pounds JG, Long GJ, Rosen JF. Cellular and molecular toxicity of lead in bone. Environ Health Perspect. 1991;91:17–32.

126. Zuscik MJ, Ma L, Buckley T, Puzas JE, Drissi H, Schwarz EM, O'Keefe RJ. Lead induces chondrogenesis and alters transforming growth factor-β and bone morphogenetic protein signaling in mesenchymal cell populations. Environ Health Perspect. 2007;115(9):1276–82.
127. Payal B, Kaur HP, Rai DV. New insight into the effects of lead modulation on antioxidant defense mechanism and trace element concentration in rat bone. Interdiscip Toxicol. 2009;2(1):18–23.
128. Long GJ, Rosen JF, Pounds JG. Cellular lead toxicity and metabolism in primary and clonal osteoblastic bone cells. Toxicol Appl Pharmacol. 1990;102:346–61.
129. Brian M. The sociology of the fluoridation controversy: a re-examination. Sociol Q. 1989;30(1):59–76.
130. Beltrán-Aguilar ED, Barker LK, Canto MT, Dye BA, Gooch BF, Griffin SO, Hyman J, Jaramillo F, Kingman A, Nowjack-Raymer R, Selwitz RH, Wu T. Surveillance for dental caries, dental sealants, tooth retention, edentulism, and enamel fluorosis—United States, 1988-1994 and 1999-2002. Centers for Disease Control and Prevention (CDC). MMWR Surveill Summ. 2005;54(3):1–43.
131. Wang W, Kong L, Zhao H, Dong R, Li J, Jia Z, et al. Thoracic ossification of ligamentum flavum caused by skeletal fluorosis. Eur Spine J. 2007;16:1119–28.
132. Bao W, Liu N, Gao B, Deng Q. Investigation of osteoarthritis in Fengjiabao, Asuo, Qiancheng villages to research the relationship between Fluorosis and Osteoarthritis. Chin J Endemiol. 2003;22(6):517–518
133. Ge XJ, et al. Investigation of osteoarthritis in middle-aged and old people living in drinking water type fluoride area in Gaomi City. Prev Med Tribune. 2006;12(1):57–58.
134. Waddington RJ, Langley MS. Altered expression of matrix metalloproteinases within mineralizing bone cells in vitro in the presence of fluoride. Connect Tissue Res. 2003;44(2):88–95.
135. Zhang XL, Xi SH, Guo XY, Cheng GY, Zhang Y. Effect of fluoride on the expression of MMP-20/TIMP-2 in ameloblast of rat incisor and the antagonistic effect of melatonin against fluorosis. Shanghai Kou Qiang Yi Xue. 2011;20(1):10–5.
136. Jowsey J, Riggs BL, Kelly PJ, Hoffmann DL. Effect of combined therapy with sodium fluoride, vitamin D and calcium in osteoporosis. Am J Med. 1972;53(1):43–9.
137. Farley JR, Wergedal JE, Baylink DJ. Fluoride directly stimulates proliferation and alkaline phosphatase activity of bone-forming cells. Science. 1983;222:330–2.
138. Cooper LF, Zhou Y, Takebe J, Guo J, Abron A, Holmen A, et al. Fluoride modification effects on osteoblast behavior and bone formation at TiO2 grit-blasted c.p. titanium endosseous implants. Biomaterials. 2006;27:926–36.
139. Monjo M, Lamolle SF, Lyngstadaas SP, Ronold HJ, Ellingsen JE. In vivo expression of osteogenic markers and bone mineral density at the surface of fluoride-modified titanium implants. Biomaterials. 2008;29:3771–80.
140. Yan D, Gurumurthy A, Wright M, Pfeiler TW, Loboa EG, Everett ET. Genetic background influences fluoride's effects on osteoclastogenesis. Bone. 2007;41:1036–44.
141. Turner CH, Garetto LP, Dunipace AJ, Zhang W, Wilson ME, Grynpas MD, Chachra D, McClintock R, Peacock M, Stookey GK. Fluoride treatment increased serum IGF-1, bone turnover, and bone mass, but not bone strength, in rabbits. Calcif Tissue Int. 1997;61:77–83.
142. Gutteridge DH, et al. A randomized trial of sodium fluoride (60 mg) +/- estrogen in postmenopausal osteoporotic vertebral fractures: increased vertebral fractures and peripheral bone loss with sodium fluoride; concurrent estrogen prevents peripheral loss, but not vertebral fractures. Osteoporos Int. 2002;13(2):158–70.
143. Everett ET. Fluoride's effects on the formation of teeth and bones, and the influence of genetics. J Dent Res. 2011;90:552–60.

Chapter 11
Biomarkers of Skeletal Muscle Injury in the Rat

Nivaldo A. Parizotto, Natalia C. Rodrigues, Paulo Sergio Bossini, and Michael R. Hamblin

Introduction

Immunohistochemistry (IHC) and immunofluorescence (IF) (sometimes both included together under the term immunocytochemistry) are techniques involving the use of antibodies for identifying proteins and molecules in cells and tissues generally viewed under a microscope. These techniques harness the power of antibodies to give highly specific recognition and binding to unique epitopes that can be short sequences of amino acids that are components of the proteins or else specific topological features of nonprotein molecules. Perhaps the most exciting part of using antibodies is that new antibodies can be generated on an as needed basis, thus providing a constant source of new reagents. Many scientists and companies are constantly generating new antibodies to specific parts of molecules, thus driving the continual expansion of IHC utilization. There have been recent efforts to generate new labels and to advance the methods of labeling molecules [1].

IHC has been used for over the last 40–45 years, initially for research but now also increasingly for diagnosis and for the assessment of specific therapeutic biomarkers [2]. In the 1980s and early 1990s, IHC was performed with a limited

N.A. Parizotto, P.T., M.Sc., Ph.D. (✉)
Physical Therapy Department, Federal University of São Carlos, São Carlos, SP, Brazil

Harvard Medical School, Boston, MA, USA
e-mail: parizoto@ufscar.br

N.C. Rodrigues, P.T., M.Sc., Ph.D. • P.S. Bossini, P.T., M.Sc., Ph.D.
Physical Therapy Department, Federal University of São Carlos, São Carlos, SP, Brazil

M.R. Hamblin, Ph.D.
Wellman Center for Photomedicine, Massachusetts General Hospital, Boston, MA, USA

Department of Dermatology, Harvard Medical School, Boston, MA, USA

Harvard-MIT Division of Health Sciences and Technology, Cambridge, MA, USA

© Springer Science+Business Media New York 2016
S.A. Aziz, R. Mehta (eds.), *Technical Aspects of Toxicological Immunohistochemistry*, DOI 10.1007/978-1-4939-1516-3_11

number of antibodies, but this has gradually changed over the last two decades to give the present availability of thousands of antibodies today. There are now many validated antibodies available for clinical use and for diagnostic use [3]. The rapid increase in immunofluorescence applications has been driven by the remarkable advances in the instrumentation for laser scanning confocal microscopy and flow cytometry that have been particularly noticeable in the last 10 years.

IHC consists of the localization of antigens on tissue sections investigated by antibodies as specific reagents through a process of antigen–antibody interaction. This antigen–antibody complex is bound by a secondary, enzyme-conjugated antibody. Many techniques can be used to amplify and visualize the staining by forming a complex with peroxidase–antiperoxidase, avidin–biotin–peroxidase, or avidin–biotin alkaline phosphatase in the tissue slide or in the cultured cells. In the presence of enzyme substrates called chromogens, the enzymes form a colored deposit at the site of antigen–antibody complex. Obviously, this color depends on the kind of chromogen that we choose. We can use either monoclonal or polyclonal antibodies to form the antigen–antibody complex, depending on their commercial availability. It is relatively simple to prepare the immunohistochemical samples to perform many kinds of analyses [4].

In the last couple of decades, there has been an exponential increase in publications on immunohistochemistry, immunofluorescence, and immunocytochemistry applications, including exploring muscle injury diagnosis. This literature is available in many cellular and molecular biology, biochemistry, pathology, histology, immunology, internal medicine, and scientific surgery articles [5]. This fact reflects the position that IHC currently holds in a pathological anatomy laboratory. It is an important tool for scientific research and also a complementary technique in the elucidation of differential diagnosis, which is not readily carried out by conventional microscopical analysis with hematoxylin and eosin and other staining methods [6].

Technical Aspects for Muscle Preparation

IHC is used to locate antigens [7] in histological sections through the antigen–antibody interaction [4]. IHC techniques are used in the detection of cell or tissue antigens ranging from proteins and nonprotein molecules to infectious agents and specific cellular populations. Especially to the muscle, it is not so different from other kinds of tissues to prepare to IHC technique. This technique consists of two phases: the first stage is the preparation of the slides (tissue fixation and processing, blocking or retrieval of binding sites (epitopes), nonspecific blocking of endogenous peroxidase, incubation with primary antibody and secondary antibody and detection enzyme, addition of chromogen, and slide assembly), and the second phase consists in the interpretation and/or quantification of the expression obtained [6].

This technique has many advantages, such as low to moderate cost, high sensitivity and high specificity for the analysis of one or more antigens simultaneously, and the need for only a small amount of muscle tissue used for examination.

Although a relatively simple technique, IHC has some particularities and the results obtained depend on several factors. A reliable result will depend on the expertise of the technician both in making and using the necessary reagents and the ability to interpret the results [8, 9]. To understand more about the structures and functions of muscle fibers and extracellular matrix, we can see reviews about this subject [10].

Brandtzaeg [7] stated that immunostaining for cell markers represents a way to "talk with cells," because it allows not only the histological origin of the cell to be identified but also indicates its function in vivo, when duly investigated with the correct antibodies. The same author affirmed that it is lamentable and non-justifiable to classify IHC as a merely descriptive method. He also emphasized that many reviewers, not aware of the accuracy of immunological detection methods, may consider them an inferior research tool whereby many manuscripts are refused for these reasons. He concluded that in vitro and in situ IHC are in fact "pictures" of the situations that occur in vivo and therefore constitute one of the pillars of biomedical research nowadays. This includes IHC, whose importance is growing gradually.

If the researchers or the pathologists look on the Internet or even in textbooks, they can find basic protocols to be used for various applications. Optimized procedures including varying the incubation time and varying the antibodies to be used and their concentrations, the species that gave rise to the antibodies, and the type of fluorophore or chromogen to be chosen may be needed to produce the best final microscopic image. A good and practical handbook for IHC was published by Burry in 2010. He gave an outline of the IHC procedures step-by-step, with a flow chart for the different preparations used to obtain the best quality results and images, including for the muscles.

Many manufacturers have in their websites suggested protocols and data on the many fluorophores, as well as monoclonal and polyclonal antibodies that are commercially available [11]. Among them may be chosen systems to analyze the required targets (proteins, specialized cellular structures, and specific cells), allowing the differentiation of the expression of immune responses as well as structural biological information.

Burry [1] suggests that it is very important to control all the steps of the process of IHC. Some questions are asked by him like "Does the antibody locations really reveal the antigen being investigated? It is easy to think that a micrograph is naturally correct, because how would the label end up in the wrong place?" For this reason, the scientists need to do appropriate controls. Control experiments are essential to rule out the possibility of a nonspecific labeling being identified as specific labeling. Recommendations for primary antibody control, secondary antibody control, and labeling control and protocols for correcting autofluorescence are available in companies that have antibodies to sell.

If the researcher plans an experiment, some decisions must choose the best methods. This also applies to IHC, for example, in determining the details to find the best reagents and preparations. The researcher must select the different applications of the methods with different types of label, then select the simplest method and a label [1]. For the muscle tissue, we can see a lot of antibodies that recognize many aspects of metabolism and/or participate in different processes suffered by the muscle cells.

Thus, there are several precautions that should be taken at all stages in the protocol of IHC, given below:

Sample fixing

Formalin-fixed samples and subsequent embedding in paraffin and subsequent sectioning to produce histological slides are used internationally [12]. But samples that are inappropriately subjected to long periods of fixation in formalin may lose their antigenicity, impairing the antigen–antibody binding. Shi et al. [13] tested a number of time periods for fixing the tissue like 6 h to 30 days; all samples showed immunostaining, but the samples fixed for 30 days were immunostained less than the samples fixed for less time. Jacobs et al. demonstrated that there is a progressive loss of antigenicity in histological sections after 12 weeks. But this response was not observed in histological sections after 10 years of storage [14].

Some researchers propose the standardization of this fixation technique, mainly to facilitate the comparison of results between different studies [12, 13]. Extensive fixation with formalin can destroy the epitopes or cause cross-linking between them [15]. Therefore, it is necessary to perform antigen unmasking or antigen retrieval, thus facilitating the binding of antibody to the target epitope. Although antigen retrieval can often be performed, some epitopes are highly sensitive to prolonged fixation in formalin and may permanently lose the ability to be unmasked or retrieved [16].

Sample processing—slices and cuts

Sections from tissue embedded in paraffin blocks must have a width between 3 and 7 μm. The sections of less than 3 μm can result in very poor immunostaining, whereas sections thicker than 7 μm can make data interpretation difficult. Many authors trying to improve the technical processing have used other media for embedding the sample, like methyl methacrylate, but it is important that the resin does not obstruct the binding of the antibodies [17]. They found that it is easy to prepare the samples with large sections to preserve the cell morphology and it is possible to remove the resin from the slices to allow binding to antibodies.

Antigen retrieval

Antigen retrieval can be done by increasing the temperature in a microwave oven or water bath, but there are antigens that require enzymatic treatment with trypsin, chymotrypsin, pepsin, or proteinase K or the combination of the two (high temperature plus enzymatic treatment) [18]. Furthermore, the duration of antigen recovery needs to be optimized in relation to different antigens as insufficient time for antigen retrieval is one of the major factors affecting the immunostaining of histological sections [2, 19].

Antibodies

Currently, there are several types of antibodies, and the variation in their dilution can change the percentage of immunostained cells [20]. As different antibodies can bind with different affinities to the same antigen, giving different results, it is necessary to know the sensitivity and specificity of each antibody [21].

Evaluation

During the preparation of a sample, an important point is the degradation of the antigen prior to fixation of the biological tissue. Fresh tissues that are subjected to prolonged drying may lose antigenicity, which can cause problems in interpretation by the researcher or the pathologist. The fixation time and the type of fixative used (the most traditional and acceptable is formalin, despite its toxic effects) can influence the antibody binding reaction [3].

Another step that is critical in sample preparation is the tissue processing. Excess of temperature of paraffin embedding (more than 60 °C) can compromise the epitope-binding sites for the binding of primary antibody [3, 22].

Immunostaining may occur, for instance, in the case of the muscle tissue, in the cell nucleus, in the cell cytoplasm, in myofibers in the muscle or in the blood vessel wall, and so on, and this will depend on the site of the target epitope of the antibody of interest. Several methods are available for evaluating the reactivity of the antibody, such as the extent of the coloration, the relative color intensity of the marking, and the combination of the intensity and extent of the color marking. All these types of evaluation require a positive control with different levels of immunostaining for comparison, and assessing the staining as mild, moderate, or strong is not the most appropriate valuation method because it will depend on the discretion of the evaluator [2].

Automated Assessment

A computerized analysis system has been used to measure immunostaining in tissues since 1980 and is the most suitable analysis to quantify the reactivity of fluorophore-labeled antibodies. However, the problem encountered with some techniques is that there is no linear relationship between the amount of antigen and the intensity of immunostaining, and as the image analysis system evaluates the marking by a measure of absorbance, if there is a nonlinear relationship, the results may not be reliable.

Standardization of IHC has been emphasized since 1977 [23], and since then, several studies have been carried out in order to optimize the standardization of this technique [24]. However, standardization remains challenging in large part, due to the presence of intrinsic uncontrollable factors, for example, variable conditions for tissue fixation and processing. These variations of processing can occur in the muscle tissue, and it can change the number of markers observed in the tissue analyses.

Tissue Characteristics

Many characteristics differentiate the kind of tissue to be used to investigate. The immunomarkers should be chosen looking for the reaction or the cell physiological mechanism involved. If the investigator knows the biochemical pathways involved in the process under investigation, the best step to be marked, and sometimes the option relating to the location of the staining on the cell structure, can be chosen.

An example here is the use of the 5-bromo-2′-deoxyuridine (BrdU) antibody to highlight the cell nucleus activity. Many researchers use this marker to differentiate if the nucleus is inside or outside of the sarcolemma in the muscle. If the nucleus is inside, it is considered a regular nucleus of the muscle cell, but, if the nucleus is outside to the membrane, the cell is determined as a satellite cell (SC), a stem cell in the muscle [25].

Muscle Tissue and IHC

In the skeletal muscle, IHC is used for the analysis of several biomarkers, such as those that report structural, metabolic, and inflammatory changes, in order to assess physiological or pathological conditions of the skeletal muscle especially after injury [10]. Conditions studied include disuse, exercise, muscular dystrophy [4], cancer [26], and aging, among others.

In our lab, we have used IHC to study the repair process in muscles. Figure 11.1 shows the muscle submitted to a cryolesion consisting of two freeze–thaw cycles in

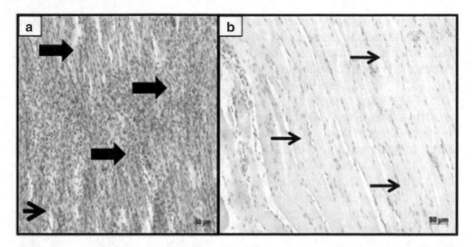

Fig. 11.1 Immunohistochemistry for COX-2 after 7 days (**a**) and 21 days (**b**) post injured muscle of rats. *Thick arrow* indicates COX-2-positive cells and *thin arrow* indicates COX-2-negative cells. Immunohistochemistry staining. *Scale bar* = 50 μm

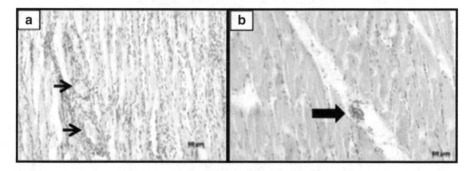

Fig. 11.2 Immunohistochemistry for VEGF after 7 days (**a**) and 21 days (**b**) post injured muscle of rats. *Thick arrow* indicates VEGF-positive cells and *thin arrow* indicates VEGF-negative cells. Immunohistochemistry staining. *Scale bar* = 50 μm

situ. Freezing was carried out by applying the flat end (0.5 × 0.5 cm) of a piece of iron and precooled in liquid nitrogen, to the surface of the proximal and distal third parts of the muscle, maintaining it in position for 10 s (twice). The left side is a muscle *tibialis anterior* of rat 7 days after the injury. COX-2 is an enzyme that is involved in prostanoid formation. It is part of the inflammation process that can be followed by IHC. There is positive COX-2 staining at 7 days post cryolesion (Fig. 11.1a). In the right side of Fig. 11.1, we can see the injured muscle after 21 days showing good muscle repair without COX-2 staining and inflammation.

In Fig. 11.2, we can see also the same time analysis after muscle injury, but the immunomarker at this time was the VEGF antibody. We can note that for this immunostaining there was an inversion of marking periods, i.e., at the early time (7 days), we cannot see specific antibody for VEGF, as expected, but 21 days after the injury (Fig. 11.2b), positive signals of VEGF marker in the muscle during repair are visible, suggesting a greater possibility of blood vessel formation which is one of the main actions of this growth factor.

Many other processes can be studied using the immunomarkers as a tool. Under a look of great interest are the SCs. Its activation and development to become a mature muscle cell such as during embryonic development, as well as the recapitulation of this process, as in muscle regeneration, have been a goal of several research groups. Knowing which cellular signals and factors affect this signaling is a key to a better understanding of the steps involved in the expression of such cells.

The quiescent SCs were identified by the expression of Pax7, a widely used marker, believed to be ubiquitously expressed in this cell population. The paired box transcription factor (Pax7) was isolated by representational difference analysis as a gene specifically expressed in cultured satellite cell-derived myoblasts. Satellite cells and muscle-derived stem cells represent distinct cell populations. Seale et al. [27] suggest that the induction of Pax7 in muscle-derived stem cells induces satellite cell specification by restricting developmental programs.

Pax7 is required for the maintenance of the postnatal skeletal muscle since the SCs are depleted in the adult *Pax7–/–* mouse. The results demonstrate Sulf1A re-expression in the regenerating muscle and rapid Sulf1A activation in vitro that precedes asynchronous MyoD activation. It is very interesting because not only an exposure to HGF but also a reduction in Sulf1A levels by neutralizing antibodies induces greatly enhanced satellite cell proliferation and a subpopulation of satellite cell progeny characterized by rapid Pax7 and MyoD downregulation without myogenin activation [28]. To achieve these goals, the authors marked with antibodies specific for Sulf1A, Pax7, and DAPI and overlapping to some of these markers. Furthermore, they performed metabolite markings that are involved with the activation and differentiation of satellite cells, including their molecular signaling. They used diverse time point markings to assess the dynamics of the process of activation and morphological organization of the muscle regeneration by SCs.

In an approach of multiple labels of proteins in the same tissue section, show unequivocal staining of the SC nucleus by MyoD and myogenin. These antibodies also mark the myoblasts and myotubes in the regenerating muscles. These authors developed a multiple marker method, sometimes in serial sections of the tissue, to compare the staining for MyoD, myogenin, c-Met, and DLK1 associated with staining for NCAN or Pax7, which are the reference markers of SCs. MyoD+ and/or myogenin+ nuclei were seen inside the basal lamina of a few dystrophin-negative myofibers, containing Pax7+ and NCAN+, Pax7– and NCAN+ cells, and some macrophage nuclei, as well as in regenerating (NCAN+) myofibers. As expected, MyoD+ and myogenin nuclei were present in fibers under repair. These multiple markers can help to identify the location of the nuclei, whether inside or outside of the basal lamina. This is one of the methods to select the SCs on muscle slides.

Horvathy et al. [29] studied repeated trauma on the muscle and their regeneration when compared with a single trauma. Using BrdU staining for newly divided cells, Hoechst staining for nucleus, and laminin staining, they could differentiate between the intact area of the muscle, the necrotic area (main lesion), and the penumbra area (the secondary lesion). The model used was cryolesion in two time points: 1 and 5 weeks after the last cold lesion. According to the results, not only muscle progenitor cells are found in the site of the lesion but also fibroblasts that can generate fibrosis and reduce the function of muscles. The authors conclude that the repeated trauma does not affect the number of proliferating cells and around 50 % of these cells are incorporated into newly formed myofibers in both single and repeated trauma groups. This indicates that single trauma and polytrauma injury does not impair the proliferating capabilities of a myogenic precursor.

In the same way, the signaling pathway of muscle trauma was analyzed by Toth et al. [30]. They explored the STAT3 signaling mediated by interleukin-6 (IL-6) as a regulator of the satellite cells. The immunohistochemical analysis revealed a progressive increase in the proportion of SCs expressing p-STAT3 with 60 % of all SCs positive for p-STAT3 at 24 h after the provocation of injury. The injury in this case was performed by 300 maximal eccentric contractions of the *quadriceps femoris* at high speed.

Although conducted in vitro, such experiments can also provide much information and learning experiences. An example of this was accomplished in the work of Wang et al. [31] when evaluating the effects of different frequencies of vibration on the formation of myotubes in C2C12 cells in culture. For this, they used the marker anti-myosin (MF-20), with Alexa Fluor 488 labeled as a secondary antibody and DAPI to mark the nuclei. They found interesting results at day 6 post treatments like the average lengths of myotubes in the 8 and 10 Hz of frequencies are significantly greater than 5 and 0 Hz groups. At day 9, the myotube lengths at all the frequencies (5, 8, and 10 Hz) were bigger than the control group (0 Hz).

IHC as a Clinical Aid for Diagnosis and Treatment Monitoring

In general, when performing experiments in animal models, the main objective is to transfer these technologies for use in diagnosis and treatment for humans. Precisely for this reason, the advances that occur in the use of antibodies in experimental animals (rats and mice in particular) can and should be carefully studied and applied to facilitate the laboratory diagnosis of muscle diseases in human subjects.

Comparative analysis of flow cytometry with IHC was conducted in human muscle samples. The IHC showed to be a powerful method to investigate the number of satellite cells using the immunomarkers NCAM and c-Met. It was reliable and effective for measuring this parameter, but the flow cytometry was faster and independent of the expertise of the investigator involved [32]. They studied the cell cycle in the same investigation using the immunomarker Pax7+, and they found that more information about the kinetics of the activities of the cell was provided when using flow cytometry compared to IHC. However, the authors do not dismiss the IHC as an effective method for this kind of analysis. The use of these kinds of markers can improve the diagnostic process for scientific investigation and in the clinical set.

The clinicians who work with patients with neuromuscular diseases know the importance of immunohistochemistry in the differentiation of dystrophies in humans, especially the recessive forms. It seems that the expression of some modified proteins is quite distinct when comparing the Xp21 dystrophy (Duchenne), Emery–Dreifuss muscular dystrophy, and the limb-girdle dystrophies. Dystrophin is the defective protein in Duchenne muscular dystrophy (DMD). Therefore, we can look at the differential labeling that is explained by gene deletions that disrupt the reading frame and give rise to the severe DMD phenotype and those that maintain it and produce the milder phenotype of Becker muscular dystrophy. Approximately 95 % of cases follow this dogma but there are several exceptions. For example, the deletion of exons 3–7 is a frameshift deletion that should result in no expression of dystrophin and a severe phenotype, but these cases often have an intermediate phenotype and some dystrophin is localized to sarcolemma [33]. The author shows other potential markers like antibodies to emerin, sarcoglycans, calpain 3, dysferlin, laminin α2, utrophin, and fast and slow isoforms of myosin that can be applied for such investigations.

In a very good review, Matos et al. [6] covered the use of IHC as an aid in surgical and clinical settings to determine the directions of treatment for patients who had biopsies. They discuss from the point of view of the pathologist and of the clinician and the surgeon the relative merits and discriminative aspects of IHC as a diagnostic tool.

The authors emphasize the importance of the method and cover possible applications. Among them may be cited: (1) histogenetic diagnosis of morphologically non-differentiated neoplasms; (2) subtype of neoplasms (e.g., lymphomas); (3) characterization of primary site of malignant neoplasms; (4) research on prognostic factors and therapeutic markers for some diseases; (5) discrimination of benign versus malignant nature of certain cell proliferations and identification of structures, organisms, and materials secreted by cells [6].

In the same review, Matos et al. [6] discuss the limitations and difficulties of IHC. They point out that the experience of the examiner may affect the preparation and interpretation of the results. However, if the procedure is carried out carefully and accurately, it can ensure very good results for later analysis. A wide variety of protocols for standardizing the IHC techniques are proposed to minimize undesirable effects. The Committee of Quality Control in IHC of the French Pathology Society published a report in 1997 demonstrating that two of the main causes of diagnostic mistakes in IHC are the nonemployment of antigen retrieval techniques and the use of amplifying methods with low power. Other renowned international quality programs are the electronic database *ImmunoQuery* (*Immunohistochemistry Literature Database Query System*) and the UK NEQAS quality program (*United Kingdom National External Quality Assessment Scheme for Immunocytochemistry*) [6]. For the diagnosis of the muscle, both for diseases, such as to detect stages of inflammation or tissue repair, it seems really important to exclude bias from the study process to achieve a high degree of assertiveness.

Concluding Remarks

As we have seen, IHC is increasingly being utilized in the research setting to study various kinds of physiological or pathological processes and in the clinical setting as to better define the muscular diagnosis and monitor treatment as collaboration between the clinician and the pathologist for muscle diseases and processes. It is important to standardize the methods to get the best quality muscle samples and analyze with more accuracy the results from the images, which needs great care and consistency, especially in the academic environment. The advent of automated image analysis software will bring changes; although it could be labor saving and increase productivity, it could also increase the likelihood of errors occurring due to lack of expertise and blindly following the computer.

Acknowledgments   MR Hamblin was supported by US NIH (grant R01AI050875). NA Parizotto was supported by FAPESP (grant 2011/06240-8) and CNPq (grant 200824/2011-2), Brazil.

References

1. Burry RW. Immunocytochemistry: a practical guide for biomedical research. New York: Springer; 2010. 223p.
2. Walker RA. Quantification of immunohistochemistry-issues concerning methods, utility and semiquantitative assessment I. Histopathology. 2006;49:406–10.
3. Yaziji H, Barry T. Diagnostic immunohistochemistry: what can go wrong? Adv Anat Pathol. 2006;13:238–46.
4. Tews DS, Goebel HH. Diagnostic immunohistochemistry in neuromuscular disorders. Histopathology. 2005;46:1–26.
5. Anagnostou VK, Welsh AW, Giltnane JM, et al. Analytic variability in immunohistochemistry biomarker studies. Cancer Epidemiol Biomarkers Prev. 2010;19:982–91.
6. Matos LL, Trufelli DC, Matos LMG, Pinhal AS. Immunohistochemistry as an important tool in biomarkers detection and clinical practice. Biomark Insights. 2010;5:9–20.
7. Brandtzaeg P. The increasing power of immunohistochemistry and immunocytochemistry. J Immunol Methods. 1998;216:49–67.
8. Jaffer S, Bleiweiss IJ. Beyond hematoxylin and eosin—the role of immunohistochemistry in surgical pathology. Cancer Invest. 2004;22:445–65.
9. Werner M, Chott A, Fabiano A, et al. Effect of formalin tissue fixation and processing on immunohistochemistry. Am J Surg Pathol. 2000;24:1016–9.
10. Chargé SBP, Rudnick MA. Cellular and molecular regulation of muscle regeneration. Physiol Rev. 2004;84:209–38.
11. Duerr JS. Immunohistochemistry, WormBook, ed. The *C. elegans* Research Community, WormBook; 25 June 2005. doi:10.1895/wormbook.1.7.1, http://www.wormbook.org.
12. Wick MR. Algorithmic immunohistologic analysis of undifferentiated neoplasms. In: United States and Canadian Academy of Pathology Annual Meeting. 1995.
13. Shi SR, Liu C, Taylor CR. Standardization of immunohistochemistry for formalin-fixed, paraffin-embedded tissue sections based on the antigen retrieval technique: from experiments to hypothesis. J Histochem Cytochem. 2007;39:741–8.
14. Manne U, Myers RB, Srivastava S, et al. Re: loss of tumor marker immunostaining intensity on stored paraffin slides of breast cancer. J Natl Cancer Inst. 1997;89:585–6.

15. Rickert RR, Maliniak RM. Intralaboratory quality assurance of immunohistochemical procedures. Recommended practices for daily application. Arch Pathol Lab Med. 1989;113:673–9.
16. Wasielewski R, Komminoth P, Werner M. Influence of fixation, antibody clones, and signal amplification on steroid receptor analysis. Breast J. 1998;44:33–40.
17. Torgersen JS, Takle H, Andersen O. Localization of mRNAs and proteins in methyl methacrylate-embedded tissues. J Histochem Cytochem. 2009;57:825–30.
18. Cattoretti G, Pileri S, Parravicini C, et al. Antigen unmasking on formalin-fixed, paraffin embedded tissue sections. J Pathol. 1993;171:83–98.
19. Rhodes A, Jasani B, Balaton AJ, et al. Study of interlaboratory reliability and reproducibility of estrogen and progesterone receptor assays in Europe. Documentation of poor reliability and identification of insufficient microwave antigen retrieval time as a major contributory element of unreliable assays. Am J Clin Pathol. 2001;115:44–58.
20. McCabe A, Dolled-Filhart M, Camp RL, Rimm DL. Automated quantitative analysis (AQA) of in situ protein expression, antibody concentration and prognosis. J Natl Cancer Inst. 2005; 97:1808–15.
21. Press MF, Hung G, Godolphin W, Slamon DJ. Sensitivity of HER-2/new antibodies in archival tissue samples: potential source of error in immunohistochemical studies of oncogene expression. Cancer Res. 1994;54:2771–7.
22. Taylor CR, Levenson RM. Quantification of immunohistochemistry-issues concerning methods, utility and semiquantitative assessment II. Hystopathology. 2006;49:411–24.
23. DeLellis RA, Sternberger LA, Mann RB, Banks PM, Nakane PK. Immunoperoxidase technics in diagnostic pathology. Report a workshop sponsored by National Cancer Institute. Am J Clin Pathol. 1979;71:483–8.
24. Taylor CR. The total test approach to standardization of immunohistochemistry (Editorial). Arch Pathol Lab Med. 2000;124:945–51.
25. Gallegly JC, Turesky NA, Strotman BA, Gurley CM, Peterson CA, Dupont-Versteegden EE. Satellite cell regulation of muscle mass is altered at old age. J Appl Physiol. 2004; 97:1082–90.
26. Alaoui-Jamali MA, Xu Y. Proteomic technology for biomarkers profiling in cancer: an update. J Zhejiang Univ Sci B. 2006;7:411–20.
27. Seale P, Sabourin LA, Girgis-Gabardo A, Mansouri A, Gruss P, Rudnicki MA. Pax7 is required for the specification of myogenic satellite cells. Cell. 2000;102:777–86.
28. Gill R, Hitchins L, Fletcher F, Dhoot GK. Sulf1A and HGF regulate satellite-cell growth. J Cell Sci. 2010;123(11):1873–83.
29. Horvathy DB, Nardai PP, Major T, Schandl K, Cselenyak A, Vacz G, Kiss L, Szendroi M, Lacza Z. Muscle regeneration is undisturbed by repeated polytraumatic injury. Eur J Trauma Emerg Surg. 2011;37:161–7.
30. Toth KG, McKay BR, De Lisio M, Little JP, Tarnopolsky MA. Parise G IL-6 induced STAT3 signalling is associated with the proliferation of human muscle satellite cells following acute muscle damage. PLoS One. 2011;6, e17392.
31. Wang CZ, Wang GJ, Ho ML, Wang YH, Yeh ML, Chen CH. Low-magnitude vertical vibration enhances myotube formation in C2C12 myoblasts. J Appl Physiol. 2010;109:840–8.
32. McKay BR, Toth KG, Tarnopolsky MA, Parise G. Satellite cell number and cell cycle kinetics in response to acute myotrauma in humans: immunohistochemistry versus flow cytometry. J Physiol. 2010;588(17):3307–20.
33. Sewry CA. Immunocytochemical analysis of human muscular dystrophy. Microsc Res Tech. 2000;48:142–54.

Index

© Springer Science+Business Media New York 2016   227
S.A. Aziz, R. Mehta (eds.), *Technical Aspects of Toxicological
Immunohistochemistry*, DOI 10.1007/978-1-4939-1516-3

Printed in the United States
By Bookmasters